Principles of Water and Wastewater Treatment Processes

Series Editor
Tom Stephenson

Volume Editor
Richard Stuetz

IWA Publishing
London • New York

Published by **IWA Publishing**
Alliance House
12 Caxton Street
London SW1H 0QS, UK
Telephone: +44 (0)20 7654 5500
Fax: +44 (0)20 654 5555
Email: publications@iwap.co.uk
Web: www.iwapublishing.com

First published 2009

Cover design by www.designforpublishing.co.uk
Typesetting by: OKS Prepress Services, India.

British Library Cataloguing in Publication Data
A CIP catalogue record for this book is available from the British Library

Library of Congress Cataloging- in-Publication Data
A catalog record for this book is available from the Library of Congress

ISBN: 1843390264
ISBN13: 9781843390268

Contents

Introduction to the *Water and Wastewater Treatment Technologies Series*

This series of texts, termed *modules*, is designed to provide an education in water and wastewater treatment from a process engineering perspective. At the end, you should have a thorough understanding of the design, operation and management of water and wastewater treatment processes. This might be the treatment of an industrial effluent to a standard acceptable for disposal to sewer, or treatment of municipal sewage to meet an environmental discharge consent, or the production of water suitable for drinking abstracted from a river. In almost all cases, a single "process", usually termed a *unit operation*, will not be able to achieve the required level of treatment. Treatment is achieved through linking the right unit operations in a *flowsheet*.

Unit operations are linked together in a flowsheet to provide water and wastewater treatment.

When designing, operating or managing a process to provide a certain quality water, it is better to first consider the individual unit operations that when linked together form the required process flowsheet for the application. This is a chemical engineering approach, as it is about the conception, development and exploitation of processes and their products. The process could be water or effluent treatment; the product a less polluted aqueous stream.

Chemical engineers use physical, chemical and biological sciences and mathematics to provide a systems approach to the understanding of changes which take place in processes, from the molecular to global scale, and to establish methods which can be employed to achieve required changes in composition, energy content, structure or physical state. In other words, chemical engineering uses a "science up" approach to solving process problems.

Understanding water and wastewater treatment through a chemical engineering approach.

This series of modules teaches the removal of pollutants from water based on just such a "science up", chemical engineering approach. To understand a unit operation, the fundamental biology, chemistry or physics underlying that unit operation needs to be properly understood. Nearly all unit operations are applicable to different levels of water and wastewater treatment, e.g. ultrafiltration membrane processes can be used in flowsheets for treating municipal and industrial effluents, as well as potable supply and pure water. The fundamental concepts of transmembrane pressure, flux rates, fouling etc., remain the same, whatever the application. Therefore the texts in this series are organised on just such a "science up" basis.

The first module, *Process Science and Engineering*, covers the chemistry, biology and chemical engineering required for a proper understanding of the unit operations covered in the main technical modules. The second module, *Principles of Water and Wastewater Treatment Processes*, introduces these unit operations, describing their overall form and application.

SERIES MODULES

- Process Science and Engineering
- Principles of Water and Wastewater Treatment Processes
- Risk Management for Water and Wastewater Utilities

Water Science at Cranfield University

Cranfield UNIVERSITY

A wholly postgraduate university, Cranfield is one of Europe's leading institutions, world-renowned for its teaching and research. Our postgraduate study programme is focused on real life rather than purely on theory. We facilitate work with industry and commerce, rather than just within the classroom. As a result of this unique approach, our graduates are found worldwide, making significant contributions in their fields.

Water Science within the School of Applied Sciences is an internationally-recognised centre of excellence undertaking research, training and consultancy in the science, engineering and management of water. Activities encompass treatment technologies, engineering, irrigation, socio-economics, management and policy as they relate to the improvement of water quality and the protection of natural, human and industrial environments.

Cranfield University has a substantive portfolio and undertakes work in developed and developing countries for industrial, government and non-governmental organizations.

Please view our web pages for more details at
http://www.cranfield.ac.uk or contact us via appliedsciences@cranfield.ac.uk

Editors

Volume Editor

Professor Richard Stuetz is Director of the University of New South Wales (UNSW) Water Research Centre (WRC), Sydney, Australia. He was previously a Senior Lecturer at Cranfield University. He has a BSc and a PhD in Environmental Biotechnology from UNSW.

Dr Stuetz is a member of the International Water Association and the Australian Water Association. He teaches water and wastewater treatment and environmental based courses to undergraduate and postgraduate students in Civil and Environmental Engineering at UNSW. His main areas of research interest are on-line instrumentation and control, fate of chemical pollutants in the environment and biological treatment processes, and the treatment and measurement of odours and volatile compounds.

Series Editor

Professor Tom Stephenson is Professor of Water Sciences and Head of School of Applied Sciences at Cranfield University. He graduated with a BSc in Biochemistry from the University of York, UK, and has a PhD in Civil Engineering from Imperial College, University of London. He joined Cranfield in 1990 and became Professor of Water Sciences in 1994. Professor Stephenson is a Chartered Engineer and a Fellow of the Institution of Chemical Engineers and the Chartered Institution of Water and Environmental Management.

Professor Stephenson has developed both full-time and part-time Masters programmes in water and wastewater process technologies at Cranfield. His research interests are centred on biological wastewater treatment.

Acknowledgements

The Series Editor and Volume Editor acknowledge the technical writing contributions of Mr Richard Hill, Visiting Fellow in Water Science and support of Water Science colleagues, Professors Judd and Parsons and Drs Cartmell and Jefferson.

How to Use This Book

This text, or module, has been designed for *individual self-paced study* i.e. you can learn in your own time without requiring additional face-to-face tuition.

KEY POINTS
As you work through each section you will find, in the margins, key points highlighted in a box.

Each section of the main text gives step-by-step learning in a particular subject. These sections are termed *Units. Aims and objectives* for each Unit are given at the start. Following these, essential prerequisites are highlighted, i.e. which other Units contain concepts that you should understand before starting on the particular Unit.

This section requires a study time of about 1 hour

At the start of each Section, in the right hand margin you will find an hourglass symbol with an approximation of how long you will need to spend on that Section. This is only an approximation, as this will be dependent on your prior knowledge and how easily you understand the subject.

The main text of each Section is enhanced with figures and tables to aid understanding. As you work through each section you will find, in the right hand margin, *key points* highlighted in boxes.

Ex 1.1
In the margin there are also exercises to help you understand the material in the main text.

In the right hand margin there are also *exercises* to help understand the material in the main text. You should attempt all exercises as you go through the text and then check your answers. These are provided at the end of each section.

At the end of the main text for each unit there are *self-assessment questions* that you should attempt to test your understanding. The answers, with working, are given at the end of the text. It is important that you do attempt the self-assessment questions *before* you look at the answers. This will help your understanding of the subject.

Introduction to Principles of Water and Wastewater Treatment Processes

The operation of unit processes in water and wastewater treatment depends on the fundamental sciences of chemistry, physics, biology and engineering. You can use the printed material to learn about the key unit operations that underpin the treatment of water and wastewater in your own time.

The text is divided into the following Units:

1. **Water Quality**
2. **Process Flowsheeting**
3. **Physical Processes**
4. **Chemical Processes**
5. **Sorption Processes**
6. **Biological Processes**
7. **Membrane Processes**
8. **Sludge Treatment and Utilisation**
9. **Odour Management**

Principles of Water and Wastewater Treatment Processes module outlines the principle unit operations that are involved in the separation, degradation and utilisation of organic and inorganic matter during water and wastewater treatment.

This module builds on the subjects of chemistry, biology and engineering covered in *Process Science and Engineering for Water and Wastewater Treatment* and provides a descriptive introduction to unit operations that are further described with numerical examples in later modules.

While every attempt has been made to ensure that all the information is correct, it is almost inevitable that there will be errors in a document of this size. If you do find any, please let us know.

1. Water Quality

Contents

Aims and Objectives

This Unit covers the principles of water quality from water and wastewater treatment works:

After studying these notes you should be able to:

1. explain in your own words the following terms and concepts:

   compliance monitoring
   performance monitoring
   process control

2. describe the different quality parameters used to monitor water quality from water and wastewater treatment works.

It is important that you are able to complete all the self assessment questions at the end of this Unit.

Essential Prerequisites

It is not necessary to have a completed any other Units before undertaking this Unit.

Additional, information on the fundamentals for this Unit, refer to the following Units in *Process Science and Engineering for Water and Wastewater Treatment*:

Unit 1 Fundamentals of Water Chemistry
Unit 4 Fundamentals of Microbiology
Unit 7 Fundamentals of Process Engineering

1.1 Introduction

Water and wastewater treatment plants are designed to provide water and wastewater to an acceptable quality, however changes to the quality and quantity of raw water and wastewater as it arrives at a treatment works as well as operational conditions, can cause changes to the quality of the water and wastewater as it is discharged from a treatment plant. These changes in water and wastewater quality can have an impact on its receptors i.e. the consumer in the case of water and the environment in the case of wastewater. Depending on the intensity of this impact, water utilities may receive different levels of response ranging from increases in the consumer complaints to prosecution by environment regulators.

To prevent the production of sub standard water and wastewater from treatment processes, water quality parameters need to be defined based on physical, organoleptic, chemical and biological measurements. As well as defining the different types of water quality parameters it is also essential to define the reasons why it is necessary to monitor for particular parameters. Different measurement parameters will provide different information on the ability of a process to handle extreme variations and still produce an acceptable product quality.

Section 1.2 requires a study time of about 1/2 hour

1.2 Monitoring

1.2.1 Compliance Monitoring

Compliance monitoring involves monitoring for a particular parameter in order to ensure that the water quality complies with a particular standard (often a legislative standard) such as potable water quality, industrial water quality or effluent discharge consent. If the water fails to meet the standard for a particular parameter we will need to know how often the limit is exceeded and by how much. Typical examples are flow, pH and the concentrations of regulated substances.

Compliance monitoring may be by on-line measurements with automatic data logging, but more often compliance monitoring involves sample collection and retrospective analysis in the laboratory, for example 90% of the UK water quality monitoring is laboratory based. Often under-rated is the fact that the logging of results provides evidence of historical patterns of compliance which, in itself, can be a valuable aid to a discharger's public relations.

ROUTINE MONITORING
Monitoring provides historical evidence of compliance and allows for optimisation of the treatment process in terms of process stability and efficiency.

1.2.2 Treatment Plant Design

The routine analysis of a water or wastewater influent to a treatment works provides an understanding of the nature of the process stream and how its quality varies seasonally or diurnally, which allows us to assess the best approach to treatment. This is particularly important in the case of the treatment of moorland surface waters, whose organic content can vary dramatically from season to season, and in industrial wastewaters from batch processes such as those used in the food and drink and pharmaceuticals industries which vary in flow and composition from hour to hour.

1.2.3 Performance Monitoring

Performance monitoring enables a water or wastewater treatment operator to determine the effectiveness or efficiency of a treatment process, in most case using indicator measurements. These measurements allow the operator to know how well a treatment process is performing in order to make any necessary adjustments to the operating conditions. Typical examples include nitrate concentrations, to check on the efficiency of nitrification in activated sludge plants and total heavy metals concentration to check on the efficiency of clarification after precipitation. These parameters are usually measured by field (Figure 1.1) or laboratory (Figure 1.2) analysis.

1.2.4 Process Control

In order to improve water and wastewater treatment efficiency, on-line monitoring of water quality parameters is often used to enable either feed forward or feedback process control. Typical examples, include raw water quality monitoring (Figure 1.3), dissolved oxygen monitoring for activated sludge aeration control, ammonia monitoring of wastewater nitrification control (Figure 1.4) and redox potential monitoring to control bisulphite addition for chromate reduction.

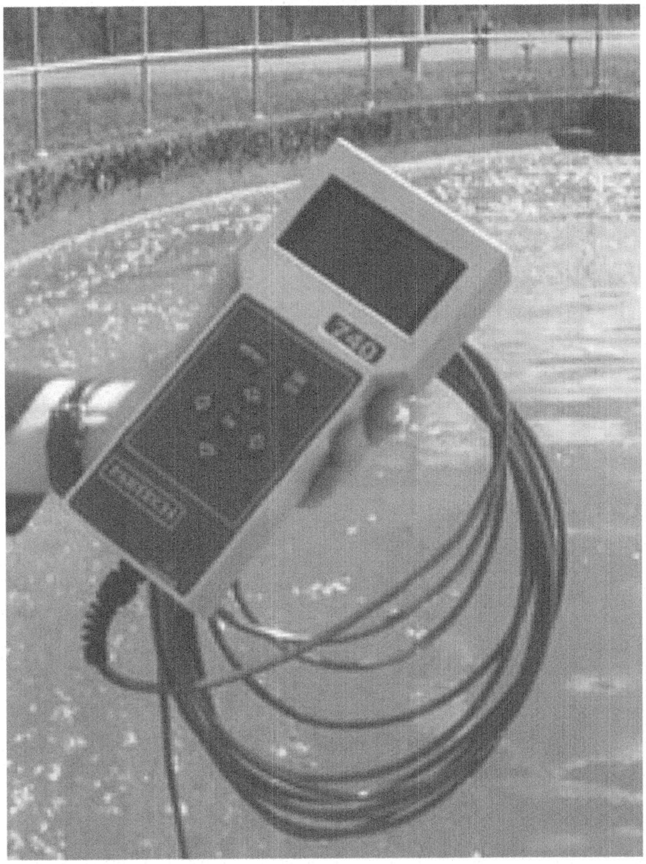

Fig. 1.1 – Portable suspended solids monitor (courtesy of Partech, UK)

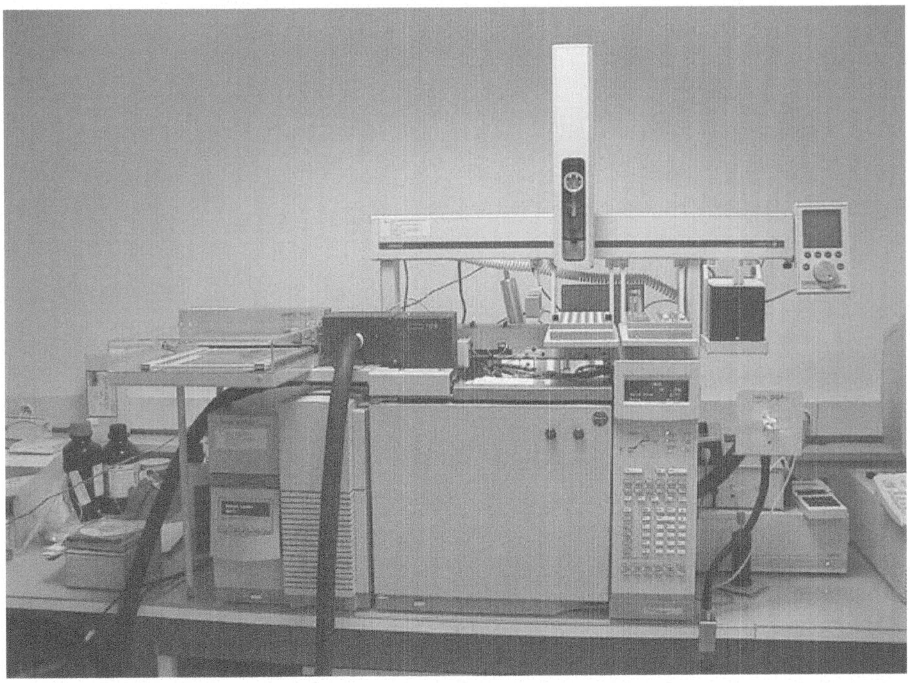

Fig. 1.2 – Laboratory based gas chromatograph mass spectrometer with olfactory port (GC-sniffer-MS) for taste and odour analysis (courtesy of Suez Environmental – CIRSEE, France)

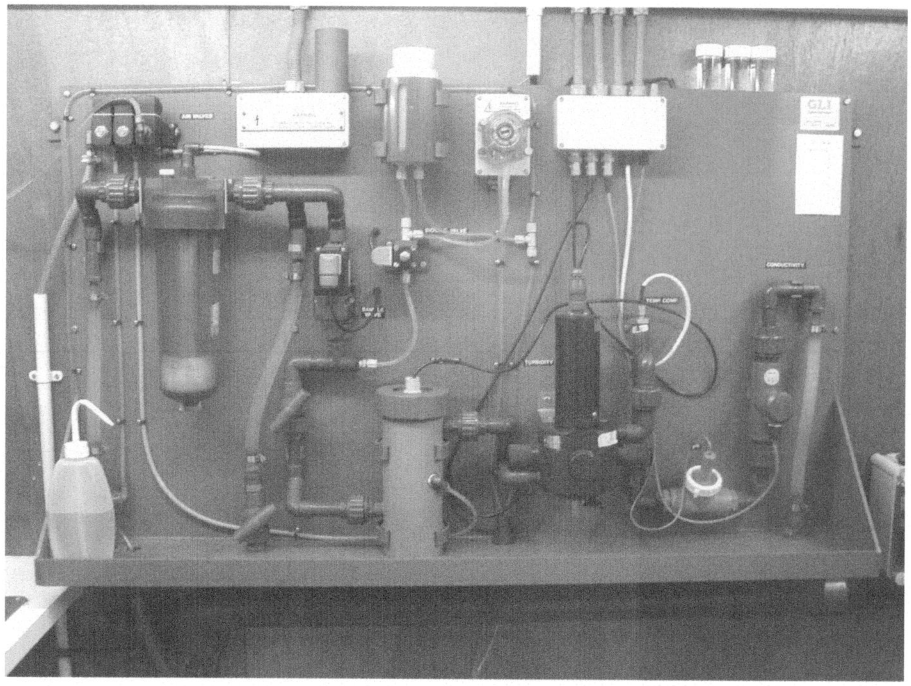

Fig. 1.3 – On-line monitoring of raw water quality parameters: colour, turbidity, iron, temperature and conductivity (courtesy of Severn Trent Water, UK)

Fig. 1.4 – In-situ ammonia monitoring of an activated sludge aeration lane (courtesy of Severn Trent Water, UK)

1.3 Measurement of Quality Parameters

Section 1.3 requires a study time of about 1/2 hour

1.3.1 Measured Parameters

It is important to know what is meant by a measured value of a parameter and how it should be interpreted. In some cases this is fairly obvious. For example, dissolved oxygen concentration is simply the concentration of oxygen dissolved in the liquid. However, it is also important to know how the determination was carried out because some tests are more sensitive than others and, in the case of laboratory determinations, may depend on the skill of the analyst. In addition many on-line measuring cells are temperature dependent and we will need to know whether this has been taken into consideration in the measurement – most instruments provide automatic temperature compensation. Of paramount importance is the statement of the units in which the parameter was measured. For example a calcium concentration of 130 mg/l expressed "as calcium" becomes 325 mg/l when expressed "as calcium carbonate" (*Process Sciences and Engineering for Water and Wastewater Treatment* – Unit 2).

Where there is a time delay between sampling and analysis – for example when a sample is sent to a laboratory – it is essential that that the sample is properly stored. Viable bacteria may deplete the dissolved oxygen (DO) concentration so that the DO value measured in the laboratory bears no relationship to that actually present on site.

In many instances the measured value has only a passing relationship with the parameter whose value we really need to know. We frequently need to know how "oxidisable" a waste stream is because we need to provide oxidation either by biological or chemical means. The available tests – Biological Oxygen Demand (BOD), Chemical Oxygen Demand (COD), Total Oxygen Demand (TOD) and Total Organic Carbon (TOC) will all give different values because of the method of analysis. When operating a biological treatment plant BOD and COD are the most important parameters but they are also the most time consuming to determine. TOD, on the other hand, is quick, simple and can be measured on-line. By carrying out parallel monitoring of COD and TOD over a period of time it may be possible to build up a relationship between these two measurements, which can allow monitoring of TOD to be used for process control instead of COD. Again absolute values are not necessary for process control purposes.

1.3.2 Accuracy and Precision

ACCURACY AND PRECISION
It is essential to be aware of the accuracy and precision of a measured parameter in order to determine its monitoring application.

In water quality monitoring, very few measurements are direct comparisons of the measured parameter against a standard as is the case in, for example, the measurement of length. Most measurements are inferential, that is some characteristic is measured and the required parameter computed. Therefore the importance of accuracy and precision is essential when discussing the measurement of parameters for water quality monitoring (*Process Sciences and Engineering for Water and Wastewater Treatment* – Unit 7). Accuracy tells us how close a set of measurements are to the real value; precision tells us how close together a set of measurements are to each other.

In the case of compliance monitoring, where analyses are essential evidence in legal proceedings, it is clear that high levels of both precision and accuracy are of vital importance. For example, wastewater flows are commonly measured by V notch weir (*Process Sciences and Engineering for Water and Wastewater Treatment* – Unit 11). In this case the height of water above the weir is measured and the flow calculated (often automatically by an electronic circuit) from this measurement. Clearly if the dimensions of the V notch are not true – for example if it has become enlarged by erosion or if it is partially blocked by weeds – then no matter how accurately the height is measured the flow computation will be incorrect.

Such an inaccuracy may be a major problem. If we are being charged a rate per unit volume of discharge then we need to know, in absolute terms, how much volume we are discharging. However, we do not always need to know the absolute value of a parameter. In performance monitoring we are usually more interested in trends, that is the value of the parameter increasing, decreasing, remaining constant or fluctuating. In this case precision is of rather more importance than accuracy. An error in measurement is acceptable provided that it is consistent in all readings. For example, an instrument that gives readings which are consistently 10% high but are within a 1% tolerance, will still be a good indicator of trends and will tell us if the plant is performing as well today as it was yesterday. An instrument which has a 5% random error and is, therefore, more accurate, will not be such a valuable tool for performance monitoring.

In most automatic process control applications, accuracy is of more importance than precision. Most automatic control loops operate by comparing the measured value with a desired value and generating a control signal which is a function of the error between the two. A large error in the measurement will, therefore, result in poor control, frequently manifested as an offset from the desired value.

1.3.3 Economics of Monitoring

Clearly it is important to carry out compliance monitoring to ensure that legislative standards are not contravened, the consequence of which might be prosecution or closure of the works. Such extreme consequences have a commercial value which is off-set by the cost of monitoring. However, there are other economic considerations for monitoring. Performance monitoring of treatment plant can lead to optimisation of operating conditions and this, in turn, allows operating costs to be minimised. Analysis of trends in waste stream quality and quantity allow water utilities to predict when new plant or processes may need to be introduced and, thus, is a useful tool in capital budgeting.

1.4 Parameter Types

Section 1.4 requires a study time of about 1/4 hour

1.4.1 Physical Parameters

Physical parameters relate to measurements which are essentially physical in nature – temperature, flow, electrical conductivity and gravimetric measurements such as the mass of suspended solids per unit volume.

1.4.2 Organoleptic Parameters

Organoleptic parameters are those measurements that are detect with our senses, that is sight – whether water is clear or *turbid* (hazy), colourless or coloured – and smell. Taste is also used in the case of potable waters where tainting compounds can cause so called *off-flavours* in drinking water supplies resulting in customer complaints.

ORGANOLEPTIC PARAMETER
Organoleptic parameter in drinking water supplies is the highest sources of consumer complaints for water utilities.

1.4.3 Chemical Parameters

Chemical parameters are those which relate to the concentrations of chemical species in water. Some of these parameters, such as cyanide, arsenic, lead, mercury and cadmium can be toxic. Others are health related: nitrates have been linked to infantile methaemoglobinaemia, nitrites to bowel cancers, sodium to heart disease. Some are measures of chemicals deliberately added such as acids and alkalis for pH control, aluminium and iron for coagulation, chlorine and ozone for disinfection, fluoride for dental health while others such as alkalinity, hardness and Total Dissolved Solids (TDS) may indicate problems if the water is used in industry for cooling or steam raising. It is necessary to understand the chemistry of water and wastewater to be able to predict how it will behave when we add or remove chemical species (*Process Sciences and Engineering for Water and Wastewater Treatment* – Unit 1).

1.4.4 Biological Parameters

Water systems include potable water, lakes, rivers and leisure pools such as swimming pools and jacuzzis. For water to be used for human consumption, it must be free from organisms that are capable of causing disease. In the late 19th century, acute waterborne diseases, such as cholera and typhoid fever were common. Since then, a whole range of infectious diseases, mainly bacterial, viral and protozoan, have been found to be water borne. It is important to remember that before a disease can occur, the host must have been in contact with an infectious dose of the disease-causing agent. This can be defined as the number of a particular pathogenic organism required to cause a disease in man and can differ for different types of organisms.

The removal of waterborne diseases is often focused on from a water treatment point of view, where reduction is through source control, filtration and disinfection, for example, chlorination. However, wastewater treatment processes can also remove a significant numbers of pathogenic organisms (Table 1.1).

PATHOGENIC ORGANISMS
Bacteria, protozoa and viruses all have the potential to cause waterborne diseases by ingestion of faecal contaminated water.

Table 1.1 – Removal of pathogenic organisms in sewage treatment

Type of Sewage Treatment	Removal Range (%)
Septic Tanks	25–75
Primary Settlement	5–40
Activated Sludge	25–99
Trickling Filters	18–99
Anaerobic Digestion	25–92
Waste Stabilisation	60–99
Tertiary (Flocculation, Sand Filtration)	93–99.99

In recent years, the number of reported illnesses relating to waterborne diseases has increased. This is mainly a result of improved diagnostic techniques for the detection of these organisms as well as an increased understanding and awareness of waterborne diseases. The range of organisms causing water-related diseases is not just limited to bacteria. Other agents include protozoa, viruses, fungi and algae.

INFECTOUS DOSE
The number of pathogenic organisms required to cause human disease.

Number varies for different pathogenic organisms.

Bacteria are isolated from water by filtration, incubation on nutrient media at 22°C and 37°C, identification by staining and enumeration by optical microscopy. Other organisms are filtered, stained where appropriate and enumerated by optical microscopy (*Process Sciences and Engineering for Water and Wastewater Treatment* – Unit 4).

The isolation, identification and enumeration of pathogenic organisms is difficult not only due to the relatively low numbers of these organisms in water but also because of difficulties in growing these organisms under laboratory conditions. Other organisms are then used as indicators of the presence of the pathogenic organisms in the water. It is critical that the *indicator organisms* must always be present when the pathogen is present and absent when the pathogen is absent. However, there are other criteria which also must be met:

- the micro-organism should only originate in the digestion tract of warm-blooded animals, especially humans;
- identification and enumeration of the micro-organism should be easy, rapid and reliable;
- the analysis should be cost effective;
- the indicator organism should survive longer than pathogens in the environment;
- the indicator organism should be present in high numbers; and
- the indicator should not be pathogenic.

INDICATOR ORGANISMS
Indicator organisms are used to detect the presence or absence of pathogenic organisms in water.

While no organism can meet all these conditions consistently, *Escherichia coli* (a faecal coliforms) fulfils most of these requirements with other coliform organisms, faecal streptococci and *Clostridium perfingens* also widely used. Two techniques are principally used to enumerate indicator organisms, membrane filtration and multiple tube methods. More recently new enzyme based techniques have emerged for microbial water testing. These tests can be used to give most probable number (MPN) values or simply indicate the presence or absence of coliforms.

1.5 Key Water Quality Parameters

Section 1.5 requires a study time of about 1 hour

1.5.1 Physical Parameters

Flow, l/s, m^3/h, Mld
Volumetric rate of liquid flow measured for compliance and process control.

Temperature, °C
Liquid stream temperature measured for compliance monitoring and process control.

Conductivity, μS/cm
The electrical conductivity of a water or wastewater which is indicative of the concentration of inorganic (ionised) salts and is used for compliance monitoring, process control and performance monitoring.

Resistivity, MΩ.cm
Reciprocal of conductivity used in compliance monitoring, process control and performance monitoring in high purity water applications.

WATER QUALITY PARAMETERS
Water quality parameters can be determined by sample collection and retrospective analysis in the laboratory and by using on-line and at-line instrumentation.

Total suspended solids (TSS), mg/l
The concentration of insoluble material in a water or wastewater measured for compliance monitoring and performance monitoring.

SDI
Silt Density Index is a measure of the rate at which a water or wastewater blocks a 0.45 μm membrane filter and is used for predicting fouling of membrane processes and performance monitoring of pre-treatment processes.

MLSS, mg/l
The suspended solids concentration in the mixed liquor in an activated sludge plant measured for performance monitoring.

Sludge density, g/100 ml
The density of sludge measured, usually, in a sedimentation tank for performance monitoring.

SVI, ml/g
Sludge Volume Index. A measurement of the comparative settleability of an activated sludge used for performance monitoring.

Ex 1.1
Which of the following is / are not measures of suspended solids concentration: TSS, MLSS, Turbidity, Resistivity and SDI?

SSVI, ml/g
Stirred Specific Volume Index. A variant of SVI.

DSVI, ml/g
Diluted Sludge Volume Index. A variant of SVI.

1.5.2 Organoleptic Parameters

Turbidity, NTU (nephelometric turbidity units), FTU (Formazin units).
A measure of the amount of colloidal material (mostly fine clay) present in a water or wastewater which produces *haze* or *cloudiness* measured by light scattering (Tyndall effect) for compliance monitoring and process control in the treatment of potable or higher quality waters.

Colour, Hazen units
A measure of the light absorbed (as opposed to scattered) as it passes through a liquid used for compliance monitoring and process control in the treatment of potable waters and industrial effluents. Colour may be due to natural vegetable tannins (humic and fulvic acids) in moorland water and to dyes in industrial wastewater.

Ex 1.2
Which organoleptic parameters are readily measurable without the use of human senses?

Tastes and Odours, *Threshold Odour Number* (TON)
Odours from drinking waters are generally described as *earthy*, *peaty*, *phenolic*, *chlorinous*. The TON is the reciprocal of the number of dilutions with odourless water which is required to make the odour undetectable by a panel of human assessors.

Visible Oil
The presence of separable oil (qv) indicated by a visible sheen on a water surface. Usually applied as a "no visible oil" limit on discharge consents.

1.5.3 Chemical Parameters (Inorganic)

CHEMICAL PARAMETERS
Chemical parameters are used for the assessment of treatment efficiency and to monitor compliance to legal standards.

pH
A measure of the acidity or alkalinity of a solution defined as $-\log_{10}[H^+]$ where $[H^+]$ represents the molar hydrogen ion concentration. The pH value ranges from 1 (very acidic) through 7 (neutral) to 14 (very alkaline) and is used for compliance and performance monitoring and process control.

Redox potential (*ORP*), mV
A measure of the voltage difference between a solution and a standard electrode representing the oxidising (+ve) or reducing (−ve) potential (and, by implication, the amount of oxidising or reducing agent present the liquid) used for process control in wastewater treatment.

TDS, mg/l
Total dissolved solids, the concentration of all materials present in solution used for compliance and performance monitoring in the treatment of water and wastewater.

Dissolved oxygen (DO), mg/l or % saturation
The concentration of oxygen gas dissolved in a liquid used for performance monitoring in aeration and deaeration processes and for process control of sewage treatment aeration systems.

Free chlorine, mg/l as Cl_2
A measure of the residual chlorine remaining in a water or wastewater after breakpoint has been achieved used for compliance and performance monitoring and process control.

Ozone, mg/l as O_3
The concentration of residual ozone present in a liquid after the ozone demand has been satisfied; used for performance monitoring and process control.

Potassium, mg/l as K
The concentration of potassium ions in a water.

Sodium, mg/l as Na or µg/l as Na
The concentration of sodium ions in a water used for compliance monitoring with potable water standards and for compliance and performance monitoring and process control in high purity water treatment.

Ex 1.3
Which of the following are chemically specific parameters: K, TDS, DO, Total Hardness, pH, Turbidity and ORP?

Total Hardness, mg/l as $CaCO_3$
The concentration of calcium and magnesium ions present in water used for performance monitoring in the treatment of industrial waters.

Iron, mg/l as Fe or µg/l as Fe
The concentration of iron in a water which may be *soluble* or *total* (soluble plus suspended) used for compliance monitoring of potable waters and performance monitoring of iron salt coagulation.

Manganese, mg/l as Mn or µg/l as Mn
The concentration of manganese in a water which may be *soluble* or *total* (soluble plus suspended) used for compliance and performance monitoring of potable waters.

Aluminium, mg/l as Al or µg/l as Al
The concentration of aluminium in a water used for compliance monitoring of potable waters and performance monitoring of aluminium salt coagulation.

Chromate, mg/l as Cr or mg/l as CrO_4^{2-}
The concentration of hexavalent chromium (chromate) ions in a water or, more usually, wastewater used for compliance and performance monitoring.

Total Chromium, mg/l as Cr
The total concentration of chromium present as both trivalent Cr^{3+} and as chromate used for compliance monitoring.

Lead, mg/l as Pb
The concentration of lead in a water (usually from corrosion of pipework) for compliance monitoring.

Mercury, mg/l as Hg
The concentration of mercury, usually from industrial wastewaters, for compliance monitoring.

Zinc, mg/l as Zn
The concentration of zinc in a water (where its presence usually indicates corrosion) or wastewater, used for compliance monitoring.

Nickel, mg/l Ni
The concentration of nickel in a water or wastewater for compliance monitoring.

Copper, mg/l as Cu
The concentration of copper in a water (where its presence usually indicates corrosion) or wastewater for compliance monitoring.

Total heavy metals, mg/l
The total concentration of all heavy metals ions in a water or wastewater for compliance and performance monitoring.

M Alkalinity, mg/l as $CaCO_3$
The concentration of bicarbonate, carbonate and hydroxide ions in a water for performance monitoring.

P Alkalinity, mg/l as $CaCO_3$
The concentration of hydroxide ions in a water or wastewater for performance monitoring in lime softening and metal precipitation processes and in boiler water conditioning.

Ex 1.4
Which of the above (Ex 1.3) directly influences the LSI?
Which measures suspended matter?

LSI
Langelier Saturation Index and the closely related *Ryznar Stability Index* are measures of the tendency of a water to deposit calcium carbonate scale. They are used in the design and performance monitoring of reverse osmosis and distillation plants and in the control of cooling water conditioning.

Chloride, mg/l as Cl
The concentration of chloride ions in a water used for compliance monitoring with potable water standards and for compliance and performance monitoring and process control in high purity water treatment.

Fluoride, mg/l as F
The concentration of fluoride ions in a water, used for compliance and performance monitoring of potable water where fluoridation is used for dental caries prophylaxis and where there is a danger of dental or osteo-fluorosis.

Bromide, mg/l as Br
The concentration of bromide ions in a water for compliance monitoring of potable waters.

Ammonia, mg/l as NH_3 or mg/l as N
Also referred to as *ammoniacal nitrogen* and as *free and saline ammonia*. The concentration of dissolved ammonia gas and ammonium ions in a water or wastewater for compliance and performance monitoring and process control.

Nitrate, mg/l as NO_3 or mg/l as N
The concentration of nitrate ions for compliance and performance monitoring of potable waters and wastewaters and process control of nitrification/ dentirification.

Nitrite, mg/l as NO_2 or mg/l as N
The concentration of nitrite ions for compliance monitoring of potable waters, performance monitoring of wastewater nitrification and chemical conditioning of cooling water circuits.

TON, mg/l as N
Total oxidised nitrogen, that is the sum of nitrites and nitrates expressed in mg/l as N for performance monitoring.

TN, mg/l as N
Total Nitrogen that is the total concentration of nitrogen from all sources – ammonia, nitrate, nitrite and organics expressed in mg/l as N for performance monitoring.

Cyanide, mg/l as CN
The concentration of cyanide ions in a water or wastewater for compliance and performance monitoring.

Phosphate, mg/l as PO_4 or mg/l as P
The concentration of phosphate ions (mostly from detergents and fertilisers) in a water or wastewater for compliance and performance monitoring.

Sulphate, mg/l as SO_4
The concentration of sulphate ions in a water or wastewater for compliance and performance monitoring.

Sulphide, mg/l as S
The concentration of sulphide ions in a water or wastewater for compliance and performance monitoring.

Silica, mg/l as SiO_2
Silica occurs as soluble silicate (or *reactive* silica) and colloidal silica. The former is usually not problematic but the latter can cause fouling of membrane processes. Silica is critical in industrial boilers and the parameter is used for performance monitoring of treatment plant.

1.5.4 Chemical Parameters (Organic)

BOD, mg/l as O_2
Biochemical Oxygen Demand is a measure of the concentration of organic matter in a wastewater which can be aerobically biologically degraded and is used for compliance and performance monitoring. The test is normally carried out over 5 days at 20°C (BOD_5), so is not suitable for process control applications.

COD, mg/l as O_2
Chemical Oxygen Demand is a measure of the concentration of organic matter in a wastewater which can be chemically oxidised by dichromates and is used for compliance and performance monitoring.

PV, mg/l as $KMnO_4$ or mg/l as O_2
Permanganate Value (also *OA, Oxygen Absorbed*) is a measure of the concentration of organic matter in a water, which can be chemically oxidised by permanganate and is used for compliance and performance monitoring.

TOD, mg/l as O_2
Total Oxygen Demand an alternative measurement of oxidisability similar to COD.

TOC, mg/l as C
Total Organic Carbon is a measure of the concentration of carbon derived from organic matter in a water or wastewater used for compliance and performance monitoring and process control.

TIC, mg/l as C
Total Inorganic Carbon is the concentration of carbon present in inorganic forms such as carbonates and bicarbonates for compliance monitoring.

VOC
Volatile Organic Carbon is the concentration of organic solvents in a water or wastewater which can be analysed by GLC and is used for compliance and performance monitoring.

TC, mg/l as C
Total Carbon is the concentration of both organic and inorganic carbon in a water or wastewater and is used for performance monitoring and process control.

Aluminoid nitrogen, mg/l as N
The total concentration of nitrogen from organic sources such as amino acids, proteins and amines in a wastewater for compliance and performance monitoring.

TKN, mg/l as N
Total Kjeldahl Nitrogen measures the concentration of nitrogen which can be biologically oxidised. It is the sum of free ammonia and organic nitrogen and is used for performance monitoring.

OKN, mg/l as N
Organic Kjeldahl Nitrogen is TKN minus free ammonia. Essentially it is the same thing as albuminoid nitrogen (qv).

Separable oil, mg/l
The concentration of oil which will float to the surface of a liquid after settlement for compliance and performance monitoring.

Total oil, mg/l
The concentration of separable and dissolved oil in a liquid for compliance and performance monitoring.

THM, μg/l
Tri-halo methanes are produced by chlorinating natural organic matter in water and are carcinogenic. The parameter is used for compliance monitoring.

Pesticides, μg/l
Pesticides and *herbicides* are toxic and/or carcinogenic and are monitored for compliance in potable water as both total pesticides and individual pesticides. Typical chemicals include Aldrin, Atrazine and Simazine.

PAH, μg/l
Polycyclic aromatic hydrocarbons are carcinogens derived from pitch linings of pipes and are monitored for potable water compliance.

Detergents, μg/l
Detergents cause taste and odour in potable water and are monitored for compliance.

Regulated Substances, μg/l
Regulated substances in discharges are toxic and/or carcinogenic and must be monitored for compliance. Lists are regularly updated but typically include chemicals like Aldrin, Atrazine, Azinphos-methyl, Cadmium, DDT, 1,2 Dichloroethane, Dichlorvos, Dieldrin, Endosulfan, Endrin, Fenitrothion, Hexachlorobenzene, Hexachlorobutadiene, Hexachlorocyclohexane, Malathion, Mercury, PCBs, Pentachlorophenol, Simazine, Trichlorobenezene, Trifluralin, Triorganotin.

1.5.5 Biological Parameters

BIOLOGICAL PARAMETERS
Biological parameters are used to detect the presence of micro-organisms that may affect treatment or water quality.

Bacteria count, cfu/ml (colony forming units) or No/ml
The most common bacterial measurement is the *heterotrophic plate count* (HPC) at 22° and 37°C but specific bacterial numbers are often required to identify specific health related infections:

- *Faecal coliforms* are found naturally in the human gut and pose no threat to health but their presence can indicate faecal contamination of a water source.
- *Salmonella,* of which there are over 2,000 known types, are all pathogenic to humans mainly causing gastroenteritis but *S. typhi* causes typhoid.
- *Shigella* is an extensive group of bacteria, the most important species being *S. sonnei* which causes the majority of waterborne infections, and *S. dysenteriae*, which causes dysentery. Polluted waters are the main source of *Shigella*-related disease, with contamination from a single identifiable source, such as a well, where disinfection is inadequate, rather than through the municipal potable water system.
- *Yersinia enterocolitica* causes acute gastroenteritis and can be carried by both humans and animals. *Yersinia* can grow at low temperatures (4°C) and has been isolated from untreated surface water frequently during colder months. It also has a marked resistance to chlorine and is hard to enumerate.
- *Campylobacter jejuni* is known to cause acute gastroenteritis for up to 1 week in humans. Other symptoms include lack of vomiting and bleeding of the gut, caused by the bacterium burrowing into gut wall.
- *Legionella pneumophila* is a common bacterium in water and soil which causes Legionnaires disease if inhaled in an aerosol so is a particular problem in atmospheric cooling towers, air conditioning systems and spray showers. It has a degree of resistance to chlorine and can colonise water distribution systems, especially if the temperature is warm.
- *Escherichia coli (E. coli)* can cause gastroenteritis in humans and is the main cause of diarrhoea in infants. The bacterium is known to be transmitted through water and its presence in natural water is indicative of faecal contamination from warm-blooded animals. More recently, this species has been linked with food poisoning though inadequately cooked beef burgers.
- *Vibrio cholerae* causes cholera: acute intestinal disease with symptoms including diarrhoea, vomiting, dehydration, and complete collapse. If no medical treatment is provided, death may occur within a few hours. *Vibrio cholera* has caused major epidemics throughout the world and can be transmitted by wastewater and polluted waters.
- *Leptospira* is carried by sewer rats and causes leptospirosis or Weils disease. This can mainly be caught through ingesting contaminated water during watersports in lakes or rivers.
- *Pseudomonas* species are opportunistic bacteria and found in potable water. They are known to have a good degree of resistance to chlorination and are one of the main causes of infections in swimming pools and whirlpools. They also can colonise high purity water systems like those in pharmaceutical factories.

Ex 1.5
Are HPC bacteria an indicator of faecal contamination?

Ex 1.6
Opportunistic bacterial pathogens threat which segment of the community?

Pyrogenicity, EU/ml (endotoxin units)
Pyrogens are *endotoxins* released when Gram negative bacteria cells break down. They induce a febrile reaction (fever) in the blood stream and must be completely absent from the high purity water used for preparing parenteral

pharmaceutical products. Endotoxins are ubiquitous in water but the low levels of toxins found in drinking water pose no great risk.

Viruses PFU/ml (plaque forming units)
Viruses are a large group of infectious agents ranging from 10 to 25 nm in diameter. They are not cells but are infectious particles that are can reproduce only inside a living cell. Water borne enteric viruses are responsible for causing infections in humans, are excreted in the faeces of infected people and may pass through sewage and water treatment processes. The most important viruses in water are:

- *Hepatitis A Virus* (HAV) causes infectious hepatitis, affecting the liver. Symptoms include fever, vomiting and diarrhoea. In extreme cases, jaundice may occur. Enteric viruses are hard to culture, and therefore, hard to detect using conventional growth methods. Recent studies show that HAV can be removed through coagulation, flocculation and filtration. HAV is also known to be more resistant to removal and inactivation than other enteric virus.
- *Norwalk-type* viruses can cause acute epidemic gastroenteritis. Like HAV, these viruses have not been cultured in the laboratory, and immunochemical methods are used to detect their presence.
- *Rotaviruses* can cause acute gastroenteritis, mainly in children and are a major cause of infant mortality in developing countries. Culturing rotaviruses in the laboratory is difficult and immunochemical techniques are used for detection. The presence of rotaviruses has been observed in drinking waters, but effective removal can be achieved during coagulation, flocculation and filtration processes.
- *Adenoviruses, Enteroviruses* and *Reoviruses* are three distinct groups of viruses that infect both the enteric and upper respiratory tract. Adenoviruses have been detected in wastewater and contaminated surface water, but not in drinking water. However enteroviruses and reoviruses have been detected in drinking water as well as wastewater and natural water. There are no proven outbreaks of illness from drinking water contaminated with these viruses, and therefore their significance as a disease-causing agent is unclear. It is possible that drinking water may be a route for the transmission of the poliomyelitis virus (an enterovirus), but this is not conclusive. Polio virus is common in effluents from biological wastewater treatment plants.
- *Bacteriophage* (or phage) is a virus that infects bacterial cells. One type of phage will only infect *E. coli*. This is called *coliphage* and will be present in water if there is faecal contamination.

VIRUSES
Viruses are infectious particles that require living host cells in order to reproduce.

Protozoan, No/ml
There are many different types of protozoan which are becoming more prevalent due to their ability to survive methods of disinfection, such as chlorination. The principal water borne protozoan are:

- *Giardia lamblia*, a flagellated protozoan which can exist for a long period of time as a cyst. When ingested, *Giardia* can cause giardiasis, a gastrointestinal disease. Symptoms include diarrhoea, fatigue and cramps and can last from a few days to months. The infectious dose for giardiasis is low, so the presence of a few cells or cysts can cause disease and the cysts can survive in water for up to 3 months. *Giardia* cysts are more resistant to chlorine than coliforms and *Giardia* cells but are much larger than bacteria and more than 99% of cysts are removed by granular media filtration. Problems can arise when the coagulant doses are low.

OOCYSTS
Oocysts are very resistant to strong oxidising biocides like chlorine.

- *Entamoeba histolytica* causes amoebic dysentery, with symptoms ranging from acute diarrhoea and fever to mild gastroenteritis. Further complications can arise if the organism invades the bloodstream and is able to reach other organs such as the liver. Cysts can be spread by the use of contaminated waters for irrigation and by using sludge as a fertilizer, and it is possible that the cysts may enter source water through this route. This protozoa is more resistant to chorine than either enteric bacteria or viruses, which may cause a problem, but the large size of the resting cysts ensures good removal by conventional filtration.
- *Cryptosporidium* occurs in many species but only two are known to cause infection in mammals. Passage of this protozoan by animals has been shown to cause infection in humans, indicating the problem of the animal reservoir and the likelihood of the entry of *Cryptosporidium* into natural and treated waters. The primary symptom of cryptosporidiosis is acute diarrhoea. Other symptoms may include abdominal pain, vomiting and fever. The infectious dose is unknown but it is thought that even 1 cell may be able to cause disease. *Cryptosporidium* is known to be ubiquitous in rivers, streams, lakes and reservoirs and higher numbers are found in waters receiving wastewater discharges. The oocysts are resistant to chlorine but both ozonation and ultraviolet irradiation have been reported to destroy them. In the UK, the Drinking Water Inspectorate requires membrane filtration where *Cryptosporidium* is present.

Algae, cells/ml

As a rule, algae are not a health concern, however some 'blue-green' algae can produce "endotoxins" or "exotoxins", which if ingested at a high enough concentration may be harmful. Under suitable conditions algae reproduce rapidly and cover streams, lakes and reservoirs in large floating *blooms*, which cause problems in water treatment and lead to taste and odour problems. Three species of blue-green algae, *Anabaena flos-aquae, Microcystis aeruginosa* and *Aphanizomenon flos-aquae*, produce exotoxins which have resulted in illness or death in mammals, birds and fish that have ingested a sufficient dose. Recreational and drinking water has caused contact irritation and possibly gastroenteritis. Blue-green algae are not directly related to true algae.

ALGAE

The production of endotoxins or exotoxins by algae can result in animal illness and/or death if ingested in high enough concentrations.

Ecotoxicity, mg/l or mg/kg

An indication of the relative danger of a waste stream discharge to aquatic life. There are various indicators including LC_{50} (concentration in mg/l of water at which 50% of test organisms are killed) and LD_{50} (dose in mg/kg at which 50% of test organisms are killed) used for performance monitoring.

1.6 Self Assessment Questions

SAQ 1.1 Give three examples of pathogenic organisms and discuss their significance in relation to waterborne diseases.

SAQ 1.2 Why are indicator organisms used for the detection of pathogens and what criteria is used for their selection?

SAQ 1.3 Define BOD and SS and discuss how this is used in relation to discharging final effluent to the environment?

SAQ 1.4 What is compliance monitoring and how does it differ from performance monitoring?

1.7 Solutions to Exercises

Ex 1.1 Resistivity

Ex 1.2 Turbidity and colour

Ex 1.3 K, DO and pH

Ex 1.4 Hardness, pH and (slightly) TDS

Turbidity

Ex 1.5 HPC bacteria are not a direct indicator of faecal contamination, however they do indicate variations in water quality and the potential for pathogen survival and regrowth.

Ex 1.6 Opportunistic bacterial pathogens do not normal threat healthy individuals, but can lead to the infection of new-born babies, the edly and immuno-compromised persons.

Bibliography

Unit 1 – Water Quality

AWWA, APHA *Standard Methods for the Examination of Water and Waste Water*. American Water Work Association and American Public Heath Association, Washington DC (2001).

Benjamin M.M. *Water Chemistry*. McGraw-Hill, New York (2002).

Fewtrell L. and Bartram J. *Water Quality: Guidelines, Standards and Health*. WHO (2001).

OECD, WHO *Assessing Microbial Safety of Drinking Water*. WHO (2003).

Smith P.G. and Scott J.S. *Dictionary of Water and Waste Management*. Butterworth-Heinemann (2002).

2. Flowsheeting

Contents

Aims and Objectives

This Unit covers the principles of flowsheeting in water and wastewater treatment:

After studying these notes you should be able to:

1. explain the principles involved in using different unit processes for the treatment of water and wastewater.
2. know the main types of unit processes used in water and wastewater treatment flowsheeting.

   primary treatment
   secondary treatment
   tertiary treatment

3. describe examples of standard water and wastewater process flowsheets

Essential Prerequisites

Prior to undertaking this Unit it is important that you have completed and understand the following Unit:

Unit 1 Water Quality

Additional, information on the fundamentals for this Unit, refer to the following Units in *Process Science and Engineering for Water and Wastewater Treatment*:

Unit 7 Fundamentals of Process Engineering
Unit 11 Engineering Hydraulics

2.1 Introduction

Water and wastewater treatment plants are a combination of separate treatment systems or *unit processes* – also termed *unit operations* – that are designed to produce an *effluent* (or final product) of specified quality from an *influent* (either water or wastewater) of known composition and quantity. The successful treatment of the different influents results from the selection of a combination of different unit processes usually in series, which is dependent upon the nature of influent. These unit processes consist of physical, chemical and biological treatment processes that either separate or treat the influent into different influent streams in order to produce a finished product of consistent quality. The quality of the final products from water and wastewater treatment plants are respectively governed by water and wastewater directives (or standards) that consist of specific water quality parameters for different chemical, physical and biological constituents.

In this unit the different unit processes used to describe the removal and treatments of either raw water or wastewater as a treatment process (or *flowsheet*) are outlined.

2.2 Flowsheeting

Section 2.2 requires a study time of about 2 hours

2.2.1 Processes

The objective of water and wastewater treatment is to produce a finished product of consistent quality that meets the water quality guidelines for supplying drinking water to consumers or discharge treat wastewater to receiving waters. A range of *unit processes* (*Process Sciences and Engineering for Water and Wastewater Treatment* – Unit 7) can be employed to produce water and wastewater that is chemically, physically and biological acceptable for the different water and wastewater industry sectors. Although these water and wastewater sectors (Table 2.1) are different in influent and effluent product quality, the *unit processes* are frequently the same. The sequence of the unit processes that are used in a water and wastewater treatment plants is defined as the *process flowsheet*.

UNIT PROCESSES
Are single treatment stages that define a specific treatment function.

Table 2.1 – Water and wastewater industry sectors

Sectors	Water types
Water supply	the influent is natural or raw water and the final product is potable water
Industrial water	the influent is either potable water from the public supply system or natural water and the product is water suitable for specific uses in industry.
Domestic wastewater	the influent is domestic sewage and the effluent must comply with consent quality allowing discharge to the appropriate receiving water.
Industrial wastewater	the influent is wastewater specific to the site and manufacturing technology used and the product must comply with consent to discharge to either municipal sewer or receiving water depending on the discharge route or for recycling and re-use within the factory.

INFLUENT AND EFFLUENT
Input is usually termed influent and effluent (more often the term used in wastewater treatment) is the output.

Each unit process has an *input* and an *output* which are defined in terms of quantity and quality:

- the flow to be treated (average, maximum and minimum and diurnal and/or seasonal variations).
- the quality of the influent (the raw water or wastewater) which means the analysis of a sample in terms of maximum, minimum and mean, but also information on how the quality varies diurnally and/or seasonally.
- the required quality of the treated water – consent limits for a waste-water discharge, quality standards for potable or industrial water reuse.

Sometimes a single unit operation will be able to convert the influent into the required effluent, but in most cases it is necessary to use a number of unit operations in series to form a *treatment process*. The input to the first unit operation is the *raw water or wastewater* which has to be treated. The output from the first unit operation becomes the input to the second unit; the product from the second becomes the input to the third and so on until the effluent from the final unit operation is the water or wastewater which the process has to produce. The selection of the appropriate unit operations to make up the treatment process is a technique which is based on an understanding of what each unit operation will do and on the experience from previous treatment process flowsheets.

The main unit processes employed in the treatment of wastewater can be classified as:

- preliminary treatment
- primary treatment
- secondary treatment
- tertiary treatment.

A process *flowsheet* is simply a sequence of the unit operations which make up a treatment process and often take the form of a *block diagram* (*Process Sciences and Engineering for Water and Wastewater Treatment* – Unit 7) like that shown in Figure 2.1.

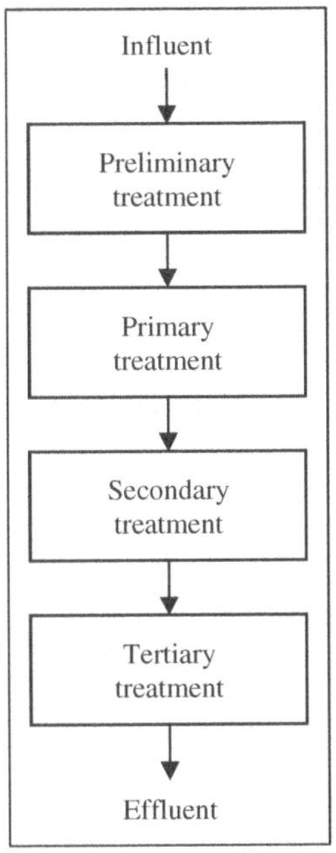

Fig. 2.1 – Summary of main unit processes for municipal wastewater treatment

FLOWSHEET
A flowsheet is a sequence of unit processes/operations that make up a treatment process.

Unit processes within a process flowsheet do not necessarily destroy the chemical and/or microbial contaminants but move them from one process to another often causing a change to their chemical form resulting in the generation of sidestreams and the need for disposal routes within process flowsheeting. The generation of waste stream raises an important question in how to deal with the disposal of different waste streams in the design and operation of water and wastewater treatment plants. Improvements in legislatives guidelines resulting in more stringent quality standards also generates the need to have continual review and updating of flowsheeting design in order to deal with these process changes and/or improved legislative standards.

2.2.2 Design Parameters

When designing a flowsheet for either a water or wastewater treatment plant, a number of the design parameters need to be considered for the whole treatment process, as well as for the separate unit operations that make up the process flowsheet. The following lists, some of the main considerations:

1. the quality of the raw water or wastewater.
2. the required effluent or potable water quality.
3. the quantity of raw water or wastewater to be treated.
4. the required quantity of the product.
5. what has to be removed.
6. the unit processes available for these various removals.
7. the order in which they are to be installed.
8. the effect of each unit process on the input quality of the next process.
9. the services, labour and utilities required.
10. the quantity and quality of waste streams produced by the unit operations (e.g. sludge production and utilisation and odour abatement).
11. the possible need for storage (e.g. between continuous and intermittent process, or to provide service water).
12. the average output of each process.
13. the actual flow rate of each process.

2.2.3 Flowsheet Selection

When designing a water and wastewater treatment plant, the order in which unit operations appear in a flowsheet varies according to the input and output design parameters for that treatment process. As a general rule, the sequence for the removal of water and wastewater quality impurities (Unit 1) can be summarised as follows:

1. removal of gross suspended solids.
2. removal of organic matter and colloids.
3. removal of suspended matter.
4. removal of dissolved organic matter.
5. removal of dissolved salts.
6. removal of dissolved gases.
7. removal of microbiological contaminants.

Usually there will be a number of alternative flowsheet designs that can be configured for a specific treatment process. The task to determine which one is the "best" often depends on selection criteria that are not design based parameters. For example flowsheets designs can have high capital cost (CAPEX) but low operating costs (OPEX) compared to those with low CAPEX and high OPEX. Factors to consider when selecting a flowsheet are:

- safety;
- capital cost (CAPEX);
- operating cost (OPEX);
- environmental impact;
- availability of chemicals;
- availability of utilities;
- quality of operating labour and availability of technical back-up;
- quality of maintenance and
- plant utilisation.

2.3 Unit Processes

2.3.1 Principles of Operation

The spectrum of unit operations used in water and wastewater treatment is wide. The simplest ones involve no more than the addition and mixing of conditioning chemicals such as:

- the addition of scale inhibiting chemicals, such as polyphosphonates and polyacrylates which prevent scale formation;
- the addition of phosphoric acid to potable water to reduce plubosol-vency;
- the addition of corrosion inhibitors such as zinc salts, molybdates and nitrites which inhibit corrosion reactions and
- the addition of biocides to control bacteria and algal proliferation.

More complex unit operations involves the interaction of multiple inputs and the production of output streams, which include effluents and sidestreams that often require further treatment and/or disposal. The operation and application of these more complex unit operations will be described in more details in the following *Units* but it is important to remember that unit operations are often used for different purposes in water and wastewater treatment. For example:

- gravity sand filters are used to filter suspended solids in water treatment and are also applied to the filtration of treated wastewater as tertiary filters;
- microfiltration and reverse osmosis membranes are used to produce potable water from sea water but they are also used in the treatment of secondary effluent to produce recycled water;
- aerobic biological treatment is used for the biodegradation of settled sewage during secondary wastewater treatment and are also applied to the removal of nitrates, iron and manganese in potable water treatment.

The main treatment processes employed in the treatment of water and wastewater can be classified according to the system changes that take place within these unit processes and can be divided into the following three categories with unit operations examples:

- Physical processes
  - Screen and bar rack
  - Grit chamber
  - Filtration
- Chemical processes
  - Coagulation
  - Flocculation
  - Disinfection
- Biological processes
  - Aerobic treatment
  - Anaerobic treatment.

Most unit operations in water and wastewater treatment result in the generation of sidestreams such as sludge, concentrated waste stream and emission of odours that often require further treatment before disposal. For example, filtration and membrane processes the effect is simply one of removing the contaminants and concentrating them into a waste stream, whereas ion exchange and chemical conditioning, actually add more chemicals into the wastewater than were already there and this may cause a problem for disposal.

The disposal of wastes from water and wastewater treatment plants is one of key design challenge in unit processes selection due to the increasing demands for more sustainable process flowsheeting. For example, in waste-water treatment, sludge treatment and final disposal is one of the ultimate considerations in evaluating different wastewater treatment design options.

The choice of unit operation also had to include economic considerations in a similar way to that for the whole flowsheet. For example, clarification can be achieved by (among other processes) sedimentation or dissolved air flotation. Dissolved air flotation has higher operating costs because it uses a pump to generate the air bubbles. On the other hand, it produces a more concentrated sludge so sludge dewatering is cheaper. It also had a higher capital cost because there is more mechanical equipment but it occupies a smaller space and the value of that space will vary depending on whether the unit operation is being constructed on a green field site or being retro-fitted into an existing crowded works. Dissolved air flotation requires more mechanical main-tenance and the availability of spares, sedimentation units can be designed to have no mechanical parts and to be virtually maintenance free.

Section 2.3 requires a study time of about 1 hour

2.3.2 Standard Unit Processes

A unit process will be placed in a water or wastewater treatment plant flowsheet to undertake a specific operation. They will require a number of services to allow for operation, often requiring an energy input, and as noted previously, there will often be sidestreams that require further treatment. Different types of unit operations will be suitable for continuous or intermittent operation. Perhaps most importantly, the overall costs, both in terms of CAPEX and OPEX, will be an important determinant in unit operation selection.

The main unit operations and their key features are shown in the boxes below. The cost of each process is shown as 'high' or 'low' to give a very broad qualitative idea of the relative costs of each. The list is not exhaustive but is designed to give an overview of the range of types of unit operations that are used in water and wastewater treatment flowsheets.

1. PROCESS:	**FLOW BALANCING**
2. WHAT DOES IT DO?	Ensures that variations in flow are smoothed out
3. FROM/TO	Usually large enough volume to ensure several hours even flow
4. SERVICES REQUIRED	None, or pumps if subsequent processes required head
5. WASTE PRODUCED	None, or solids as it may act as sedimentation
6. COST	**LOW**
7. OPERATION	**INTERMITTENT**
8. NOTES	Danger of biological activity in strong wastes

1. PROCESS:	**DEGASSING**
2. WHAT DOES IT DO?	Removing dissolved carbon dioxide
3. FROM/TO	From 250 to 5 mg/l typically
4. SERVICES REQUIRED	Air blower and repump
5. WASTE PRODUCED	None
6. COST	**LOW**
7. OPERATION	**CONTINUOUS**
8. NOTES	Particulants and bacteria blown into water from atmosphere.

1. PROCESS:	**GRIT REMOVAL**
2. WHAT DOES IT DO?	Dense particles (c 2500 kg/m^3) taken out
3. FROM/TO	Almost complete removal, without removing less dense solids
4. SERVICES REQUIRED	Small amounts of electricity for degritting
5. WASTE PRODUCED	Grit, high density, low in biological solids
6. COST	**LOW**
7. OPERATION	**CONTINUOUS**
8. NOTES	Relies on gravity, like sedimentation, but for denser particles

1. PROCESS:	**REVERSE OSMOSIS**
2. WHAT DOES IT DO?	Removes dissolved solids (ions) across a membrane by pressure. RO removes all colloids and particulates
3. FROM/TO	From c 50,000 mg/l down to 10–100 mg/l, lower in special applications
4. SERVICES REQUIRED	Electricity (for pressure, or direct) membrane cleaning system, chemical dosing and cleaning
5. WASTE PRODUCED	Concentrated solution, but RO particularly be very wasteful of water
6. COST	**HIGH**
7. OPERATION	**CONTINUOUS**
8. NOTES	Membrane fouling the main problem

1. PROCESS:	**SCREENING**
2. WHAT DOES IT DO?	Removes particles and floating objects using bars or mesh
3. FROM/TO	Incoming objects >1 mm, at lowest, removed completely
4. SERVICES REQUIRED	None, unless automatic descreening
5. WASTE PRODUCED	Screenings, containing some biological solids, rags, plastics, plants etc. Low volume compared to total flow
6. COST	**HIGH/LOW**
7. OPERATION	**CONTINUOUS/INTERMITTENT**
8. NOTES	Screenings often need dewatering

1. PROCESS:	**PRECIPITATION**
2. WHAT DOES IT DO?	Removes some soluble ions by addition of chemical – especially metals
3. FROM/TO	Can reduce anions and cations from 100's mg/l to c 10 mg/l
4. SERVICES REQUIRED	Chemical supply tanks, power for mixing. Lime commonly used and presents handling problems
5. WASTE PRODUCED	An inorganic sludge or slurry
6. COST	**LOW**
7. OPERATION	**INTERMITTENT**
8. NOTES	Many different chemicals for different removal processes. Sludge disposal may be difficult

1. PROCESS:	**ELECTRODIALYSIS**
2. WHAT DOES IT DO?	Removes dissolved solids (ions) across a membrane by electricity
3. FROM/TO	From c 50,000 mg/l down to 10–100 mg/l, lower in special applications
4. SERVICES REQUIRED	Electricity, membrane cleaning system, chemical dosing and cleaning
5. WASTE PRODUCED	Concentrated solution
6. COST	**HIGH**
7. OPERATION	**CONTINUOUS**
8. NOTES	Membrane fouling the main problem

1. PROCESS:	SEDIMENTATION
2. WHAT DOES IT DO?	Removes heavier-than-water particles by settlement
3. FROM/TO	Up to perhaps 5,000 mg/l of solids. Product widely dependent on particle size and density, but typically contains between 1 to 5% of incoming concentration
4. SERVICES REQUIRED	None, except electricity where mechanical agitation is used
5. WASTE PRODUCED	Sludge, which may be as low as 1% solids, or too high to be readily pumpable, depending on nature of solids
6. COST	LOW
7. OPERATION	CONTINUOUS
8. NOTES	In most cases requires long residence time and so has large footprint. May need filtration downstream

1. PROCESS:	ION EXCHANGE
2. WHAT DOES IT DO?	Removes IONS (cations and anions) by transfer to a solid material in exchange for an equivalent number of another ion
3. FROM/TO	Incoming ions in 100's of mg/l down to any level desired – capable of producing 'ultra pure' water
4. SERVICES REQUIRED	Regeneration chemicals, low pressure air
5. WASTE PRODUCED	Solution high in removed ions when exchanger is regenerated
6. COST	HIGH
7. OPERATION	INTERMITTENT
8. NOTES	Vast variety of ion exchange materials and process flow sheet options

1. PROCESS:	THICKENING
2. WHAT DOES IT DO?	Increases the solid content of sludges
3. FROM/TO	Solids content increased by c 50%, starting at 4–8% w/v
4. SERVICES REQUIRED	Some mechanical power or air, depending on technique
5. WASTE PRODUCED	Supernatant liquor, very high in dissolved solids
6. COST	LOW
7. OPERATION	INTERMITTENT
8. NOTES	'Picket fence' thickeners used for biological sludge

1. PROCESS:	**DEWATERING**
2. WHAT DOES IT DO?	Increases the solid content of sludges
3. FROM/TO	Solids content increased from c 5% to 25–50%
4. SERVICES REQUIRED	Power pumps, pressure in filters, motors
5. WASTE PRODUCED	Supernatant liquor, high in dissolved solids
6. COST	**HIGH**
7. OPERATION	**INTERMITTENT**
8. NOTES	Output sludge difficult to handle. Includes cetrifuges, belt presses

1. PROCESS:	**FLOTATION**
2. WHAT DOES IT DO?	Fine gas bubbles entrap solids and remove them at surface
3. FROM/TO	Clear water: after flocculation to near clear, Dirty water: after biological stage to near clear
4. SERVICES REQUIRED	Compressed air or vacuum for fine bubble
5. WASTE PRODUCED	Sludge of approx 2–15% solids
6. COST	**HIGH**
7. OPERATION	**CONTINUOUS**
8. NOTES	Low space requirements. Quick start-up. Alternative to sedimentation (for non-settling suspended particles)

1. PROCESS:	**FLOCCULATION**
2. WHAT DOES IT DO?	Increase the size of fine particles
3. FROM/TO	Unsettleable particles to settling velocities of c 1.5 m/h
4. SERVICES REQUIRED	Chemical dosing equipment, power for rapid mixing but must avoid high shear
5. WASTE PRODUCED	None at this stage
6. COST	**LOW**
7. OPERATION	**CONTINUOUS**
8. NOTES	Possible with help of polymer as well as coagulate chemical

1. PROCESS:	**UV DISINFECTION**
2. WHAT DOES IT DO?	Destroys micro-organisms, but almost never completely
3. FROM/TO	Several log reduction of bacteria numbers
4. SERVICES REQUIRED	Electricity
5. WASTE PRODUCED	None
6. COST	**HIGH**
7. OPERATION	**CONTINUOUS**
8. NOTES	Amount of disinfection dependent on power input. Presence of particulates reduces efficiency

1. PROCESS:	**CHLORINATION DISINFECTION**
2. WHAT DOES IT DO?	Destroys micro-organisms, but almost never completely
3. FROM/TO	Depends on input. If low, near-complete removal can be achieved. If high (< 1,000 bacteria per ml), 2–3 log reductions can be achieved
4. SERVICES REQUIRED	Chemicals for dosing (chlorine gas etc.)
5. WASTE PRODUCED	Chlorinated organic compounds
6. COST	**HIGH**
7. OPERATION	**CONTINUOUS/INTERMITTENT**
8. NOTES	If enough chlorine is added, prevention of microbial growth after this unit operation can be achieved. Precautions in handling toxic $C1_2$

1. PROCESS:	**AEROBIC BIOLOGICAL FIXED FILM**
2. WHAT DOES IT DO?	Degrades dissolved and suspended organic compounds in presence of oxygen
3. FROM/TO	Up to 95% removal of dissolved organic compounds at influent concentrations c 100–1000 mg/l BOD
4. SERVICES REQUIRED	None usually, some electricity
5. WASTE PRODUCED	Biological solids ('humus') that falls off the support, carbon dioxide
6. COST	**LOW**
7. OPERATION	**CONTINUOUS/INTERMITTENT**
8. NOTES	Micro-organisms grown on the surface of an inert material, e.g. stone or plastic. Requires large land areas

1. PROCESS:	**AEROBIC BIOLOGICAL SUSPENDED GROWTH**
2. WHAT DOES IT DO?	Degrades dissolved and suspended organic compounds in presence of oxygen
3. FROM/TO	Up to 95% removal of dissolved organic compounds, influent BOD 100–10,000 mg/l. Suspended solids increases from c 100 to 2000 mg/l
4. SERVICES REQUIRED	Electricity for aeration
5. WASTE PRODUCED	Biological solids c 2000 mg/l, i.e. the suspended solids, carbon dioxide
6. COST	**HIGH**
7. OPERATION	**CONTINUOUS**
8. NOTES	Process also able to removal metals etc. by adsorption. Can also remove ammonia

1. PROCESS:	**AEROBIC BIOLOGICAL SUSPENDED GROWTH**
2. WHAT DOES IT DO?	Degrades dissolved and suspended organic compounds in presence of oxygen
3. FROM/TO	Up to 95% removal of dissolved organic compounds, influent BOD 100–10,000 mg/l. Suspended solids increases from c 100 to 2000 mg/l
4. SERVICES REQUIRED	Electricity for aeration
5. WASTE PRODUCED	Biological solids c 2000 mg/l, i.e. the suspended solids, carbon dioxide
6. COST	**HIGH**
7. OPERATION	**CONTINUOUS**
8. NOTES	Process also able to removal metals etc. by adsorption. Can also remove ammonia

1. PROCESS:	**ANAEROBIC BIOLOGICAL FIXED FILM**
2. WHAT DOES IT DO?	Removes organic material in absence of oxygen
3. FROM/TO	c 70% removal of high organic concentrations, i.e. 10,000 mg/l
4. SERVICES REQUIRED	Heating, often via methane produced by the process
5. WASTE PRODUCED	Biological solids, methane, carbon dioxide
6. COST	**HIGH**
7. OPERATION	**CONTINUOUS**
8. NOTES	Influent should be low in solids

1. PROCESS:	**ANAEROBIC BIOLOGICAL SUSPENDED GROWTH**
2. WHAT DOES IT DO?	Reduces biological solids and organics by microbes grown in absence of oxygen
3. FROM/TO	Halves organic solids from c 4–8% w/v. As a consequence, reduces BOD
4. SERVICES REQUIRED	Heating, often via methane produced. Energy for mixing
5. WASTE PRODUCED	Biological solids, methane, carbon dioxide
6. COST	**LOW**
7. OPERATION	**INTERMITTENT**
8. NOTES	Bacteria grow very slowly, so relatively large reactors needed

1. PROCESS:	**COAGULATION**
2. WHAT DOES IT DO?	Precipitation of Al or Fe creates fine flocs which absorbs suspended and large dissolved impurities
3. FROM/TO	Best on concentration less than say 50 mg/l impurity be removed: removal can be almost complete
4. SERVICES REQUIRED	Alum or ferric solution dosing gear
5. WASTE PRODUCED	Suspended precipitate must be removed
6. COST	**HIGH/LOW**
7. OPERATION	**CONTINUOUS/INTERMITTENT**
8. NOTES	Difficult to control, especially on variable water, and therefore unsuitable for small throughputs. Alternative to flotation

1. PROCESS:	**ULTRAFILTRATION**
2. WHAT DOES IT DO?	Small pore size (<0.1 μm) so can remove colloids, high molecular weight organics and particles
3. FROM/TO	Complete removal of particulates in low concentrations, e.g. bacteria 100–1000 per ml. Turbid solutions from c 20 NTU to zero
4. SERVICES REQUIRED	Pressure pumping (usually less than 5 bar), cleaning chemicals
5. WASTE PRODUCED	Solids when backwashed
6. COST	**HIGH**
7. OPERATION	**INTERMITTENT**
8. NOTES	Fouling a potential problem, cleaning routine the key to successful operation

1. PROCESS:	**DEPTH FILTRATION**
2. WHAT DOES IT DO?	Removes heavier than water particles by filter action. Does not removal particles less than c 3 μm
3. FROM/TO	Applicable for removal of suspended solids influent at 5 to 50 mg/l down to <1 NTU
4. SERVICES REQUIRED	High flow-rate pumps or high level storage for backwash water and low pressure air
5. WASTE PRODUCED	Sludge, containing organic and inorganic material, usually less than 1% solids
6. COST	**LOW**
7. OPERATION	**INTERMITTENT**
8. NOTES	Inert material used in filter e.g. sand, or sand and anthracite

1. PROCESS:	**ACTIVATED CARBON**
2. WHAT DOES IT DO?	Removes non degradable organics by adsorption
3. FROM/TO	From mg/l and μm/l levels to ng/l levels
4. SERVICES REQUIRED	High temperature for regeneration of carbon
5. WASTE PRODUCED	Carbon can be regenerated by heat
6. COST	**HIGH**
7. OPERATION	**CONTINUOUS**
8. NOTES	Regeneration often performed off-site

1. PROCESS:	**OZONATION**
2. WHAT DOES IT DO?	Removes micro-organisms and oxidises chlorinated organic compounds
3. FROM/TO	Bacterial counts from 100's to near zero, organics from μg/l to ng/l levels
4. SERVICES REQUIRED	Electricity if on-site ozone produced
5. WASTE PRODUCED	Leaves some oxidised organic compounds in water
6. COST	**HIGH**
7. OPERATION	**CONTINUOUS**
8. NOTES	Precautions in handling toxic 0_3

1. PROCESS:	**CROSSFLOW MICROFILTRATION**
2. WHAT DOES IT DO?	Removes suspended solids and bacteria for particle sizes $<\sim 1\ \mu m$
3. FROM/TO	Almost complete removal or particulates in moderate to high concentrations
4. SERVICES REQUIRED	Pressure pumping (usually $\sim$1 bar), may use chemicals for cleaning or membrane formation
5. WASTE PRODUCED	Solids when backwashed
6. COST	**HIGH**
7. OPERATION	**INTERMITTENT**
8. NOTES	Fouling a potential problem, cleaning routine the key to successful operation. Small footprint

SAQ
For each unit process define the main principle of operation i.e. one of:
- biological
- chemical
- physical.

2.4 Flowsheets

FLOWSHEET
A flowsheet is a sequence of unit operations that make up a treatment process

The basic unit operations can be put together to form a flowsheet to transform an input (influent) into a desired output (effluent).

Figures 2.2 and 2.3 show two flowsheets for potable water treatment; both would be for treating surface waters before putting into public supply. In contrast, it is possible to take borehole water from an aquifer and merely have chlorine disinfection before putting into the water supply network. For raw waters that come from pristine catchments, where there will be little agricultural activity and associated pollutant in runoff, a flowsheet that removes solid and colloidal material through coagulation-flocculation treatment and depth filtration should suffice (Figure 2.2). Where the catchment has received inputs from runoff containing materials such as organic matter and pesticides, additional unit operations will be included in the flowsheet for their removal (Figure 2.3). Chemical oxidation through ozonation will transform recalcitrant organic pollutants and activated carbon will adsorb others. For the flowsheet shown in Figure 2.3, dissolved air floatation has been used to remove the larger particulates instead of coagulation-flocculation. In both flowsheets, chlorination is used to provide disinfection prior to putting into the public water supply network.

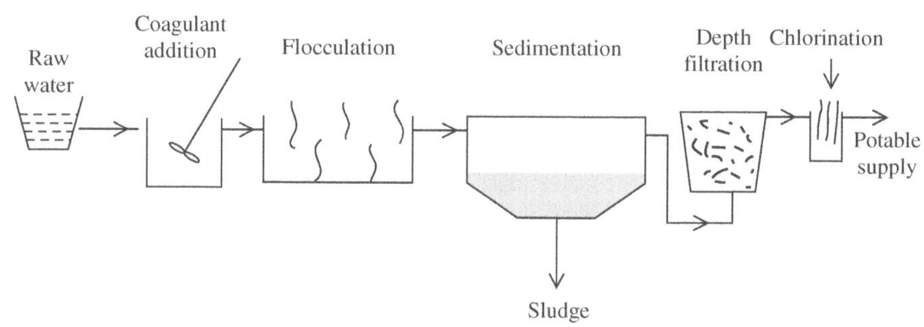

Fig. 2.2 – Flowsheet for potable water treatment

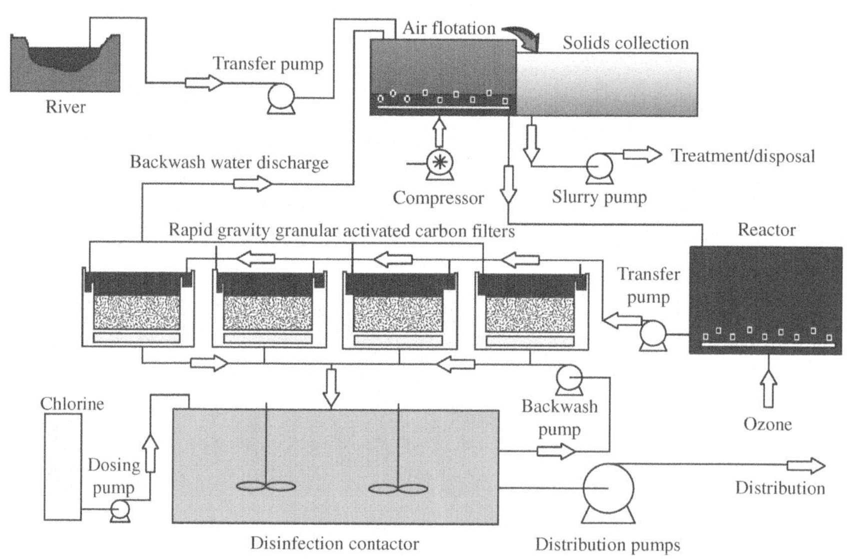

Fig. 2.3 – Flowsheet for advanced potable water treatment

Where high purity water is required, a typical flowsheet might include a series of physical unit operations designed to remove all particulate matter along with almost all dissolved organic and inorganic matter (Figure 2.4). 'Pyrogens' are the remnants of bacteria that will be found in tapwater and these are a key contaminant that needs removing from potable quality water prior to use in pharmaceutical production. Membrane unit processes that consume relatively large amounts of energy will be key to providing the right quality product water.

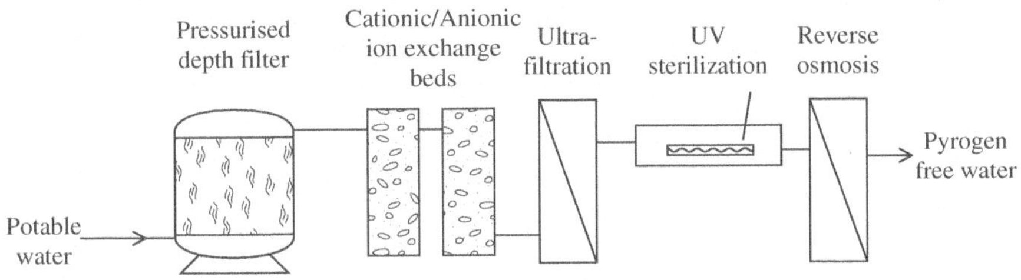

Fig. 2.4 – Flowsheet for production of pyrogen-free water from potable water

In municipal wastewater treatment, sedimentation unit operations play a crucial role and may appear at several points in the flowsheet (Figures 2.5–2.8). Central to the flowsheet will be a biological process designed to remove organic matter. The first process designed for such duty is the trickling filter, still deployed in many countries (Figure 2.5). More commonly, the activated sludge process will be used; this process offers a smaller footprint and greater control (Figure 2.6). However, the other components in the flowsheet will be the same as for a trickling filter. Where a better quality effluent is needed, the biological process will be adapted, e.g. to remove nitrate nitrogen through addition of anoxic zones to the activated sludge process, and tertiary process may be added, e.g. reed beds, to provide an effluent with lower suspended solids and organic matter concentrations (Figure 2.7). Figure 2.8 shows the application of a more recently developed biological unit operation, the membrane bioreactor, to provide final effluent suitable for discharge to receiving waters that require low microbiological contamination.

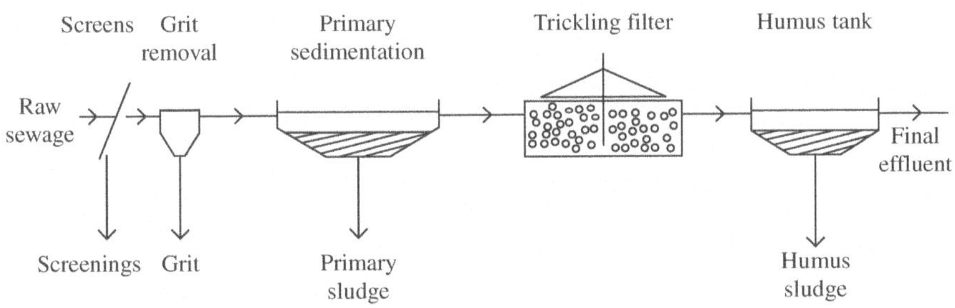

Fig. 2.5 – Flowsheet for municipal sewage treatment using trickling filters

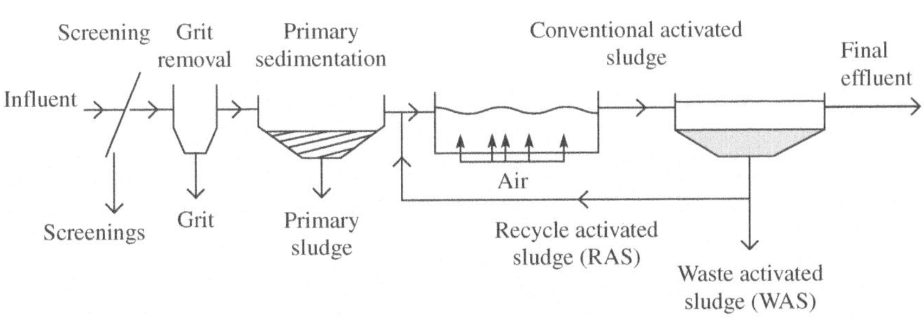

Fig. 2.6 – Flowsheet for standard municipal wastewater treatment

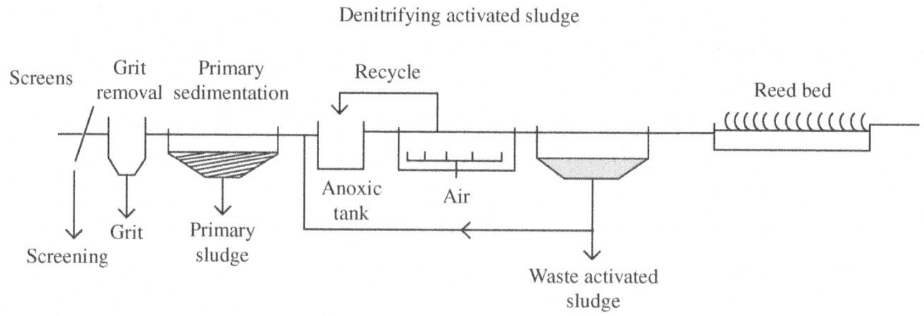

Fig. 2.7 – Flowsheet for municipal wastewater treatment with total nitrogen removal

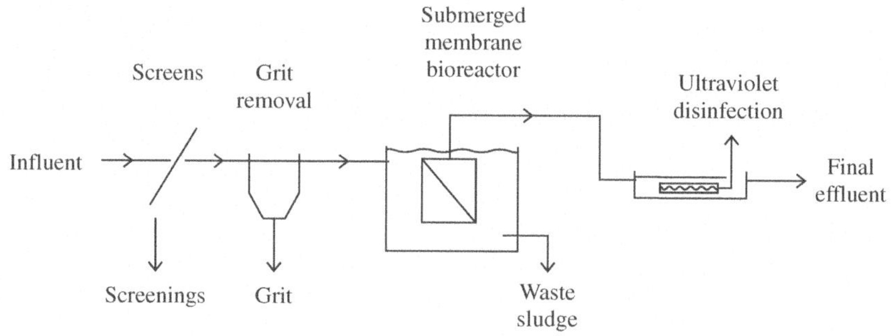

Fig. 2.8 – Flowsheet for municipal sewage treatment to provide effluent meeting strict disinfection standards

In the flowsheets shown in Figures 2.2 to 2.8, there are a number of sidestream flows. Often these will be higher solids concentration flows i.e. sludges. Where they arise from physio-chemical processes, they can mainly consist of inorganic material, but also contain organic solids. In wastewater treatment, the sidestreams are mainly organic sludges. These will require further treatment and an illustrative flowsheet is shown in Figure 2.9.

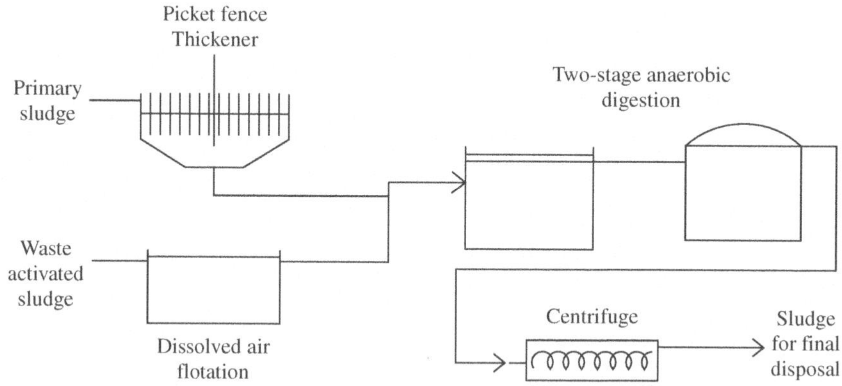

Fig. 2.9 – Outline flowsheet for municipal sludge treatment

In physical appearance, unit operations may not necessarily appear as separate entities. For example, a package sewage treatment plant, designed for relatively small flows, may incorporate screening, primary sedimentation, a secondary biological process and final sedimentation within a single structure.

2.5 Self Assessment Question

Main principles of operation for unit operations

B = Biological, C = Chemical, P = Physical

Flow Balancing	P
Degassing	P
Grit Removal	P
Reverse Osmosis	P
Screening	P
Precipitation	C
Electrodialysis	C
Sedimentation	P
Ion Exchange	C
Thickening	P
Dewatering	P
Flotation	P
Flocculation	C
UV Disinfection	C
Chlorination	C
Aerobic Biological Fixed Film	B
Aerobic Biological Suspended Growth	B
Anaerobic Biological Fixed Film	B
Anaerobic Biological Suspended Growth	B
Coagulation	C
Ultrafiltration	P
Depth Filtration	P
Activated Carbon	C
Ozonation	C
Crossflow Microfiltration	P

Bibliography

Unit 2 – Flowsheeting

Droste RI.L. *Theory and Practice of Water and Wastewater Treatment*. John Wiley & Sons Inc., Chichester (1997).

Gray N.F. *Water Technology*. Arnold, London (1999).

Metcalf and Eddy. *Wastewater Engineering, Treatment and Reuse*. 4th ed., McGraw Hill, New York (2003).

Parsons S.A. and Jefferson B., *Introduction to Potable Water Treatment Processes*, Wiley-Blackwell (2006).

3. Physical Processes

Contents

Aims and Objectives

This Unit covers the principles of physical processes used in water and wastewater treatment:

After studying these notes you should be able to:

1. explain in your own words the following terms and concepts:

   sedimentation
   screening
   filtration

2. know the main types of physical processes

   sedimentation tanks
   clarification tanks
   bar screens
   travelling screens
   deep bed filtration
   centrifugation

It is important that you are able to complete all the self assessment questions at the end of this Unit.

Essential Prerequisites

It is not necessary to have a completed any other Units before undertaking this Unit.

Additional, information on the fundamentals for this Unit, refer to the following Units in *Process Science and Engineering for Water and Wastewater Treatment*:

Unit 7 Fundamentals of Process Engineering
Unit 11 Engineering Hydraulics
Unit 12 Particle Sedimentation

3.1 Introduction

Many of the impurities in water and wastewater are in the form of suspended particles. Gross solids to be removed from aqueous streams can be categorised as follows:

- large floating and suspended solid objects e.g. pieces of wood, vegetable debris, fish, animal carcasses, plants, plastic debris, paper and rags;
- silt and grit, typically particles of 10–1000 μm diameter with density 2000 2600 kg/m^3 and
- flocs and biomass, typically 100–1000 μm diameter with density 1005–1010 kg/m^3.

In order to provide effective removal of these different size particles during water and wastewater treatment a range of physical processes can be employed such as screening, grit removal, sedimentation, flotation and filtration.

In this unit the principle physical processes used to remove gross solids in order to protect downstream equipment and processes are outlined along with physical process designed to remove solids with densities a little greater than that of water.

Section 3.2 requires a study time of about 1 hour

3.2 Sedimentation

Sedimentation is the settlement of particles which have a higher density than the liquid in which they are suspended under the influence of gravity (*Process Sciences and Engineering for Water and Wastewater Treatment* – Unit 3).

Simple settlement tanks are widely used for the preliminary treatment of very high suspended solids waters and for the storage of wastewater prior to wastewater treatment such as:

- removal of grit from domestic sewage (where they are usually called *grit channels*);
- storage of raw sewage during storm conditions (usually called *storm tanks,* Figure 3.1) and
- balancing of intermediate wastewater flows.

SETTLING TANKS
Settling tanks remove suspended particles by gravity separation.

Vertical and radial flow designs are employed for lower concentrations of lower density solids including:

- removal of large solids in wastewater primary treatment;
- removal of flocs in water clarification;
- removal of lime softening precipitates;
- removal of precipitated metals from industrial effluents and
- removal of biomass in biological wastewater clarification.

Fig. 3.1 – Wastewater storm tanks (courtesy of Anglian Water, UK)

3.2.1 Horizontal Settlement Tanks

Horizontal *settling tanks* are quite simple rectangular tanks typically about 2m deep with a length: width ratio of about 2–5 (Figure 3.2). Water enters at one end and leaves via an overflow weir at the other. Solids settle to the bottom and the sludge is usually mechanically scraped to one end by a *chain and flight scraper* for removal (Figure 3.3).

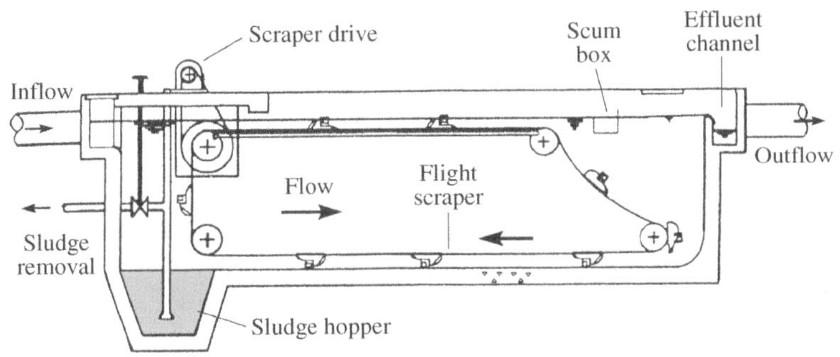

Fig. 3.2 – Horizontal settling tank (after Gary, 1999)

Fig. 3.3 – Horizontal settling tank showing scraper (courtesy of Huber, Germany)

Constant velocity channels are conceptually the simplest system for removing grit (Figure 3.4). These channels are either parabolic or trapezoidal in cross section and designed to give a constant liquid flow velocity of about 0.3 m/s under variable flow conditions. Alternative square grit removal chambers or *detritors*, the flow enters via baffles for distribution and a scraper moves the grit into a sump for periodic removal.

Fig. 3.4 – Constant velocity channels (courtesy of Anglian Water, UK)

Sedimentation plants are designed to remove particles of a given size and density so their performance depends very much on the size distribution of the influent suspended solids. They can usually be expected to remove 90% of suspended solids of the design size or larger.

3.2.2 Vertical and Radial Settlement Tanks

Vertical flow (Figure 3.5) and radial flow settling tanks (Figure 3.6) are commonly used for primary settlement of screened sewage, because they take up less space than horizontal tanks. Raw water enters via a *diffuser drum* in the centre which directs the flow downwards to the bottom of the tank. As the liquid flows upwards in the case of a vertical flow tank or outwards in the case of radial flow, the liquid velocity reduces allowing solids to settle.

VERTICAL FLOW TANKS
As the liquid flows upwards the liquid velocity reduces allowing the solids to settle.

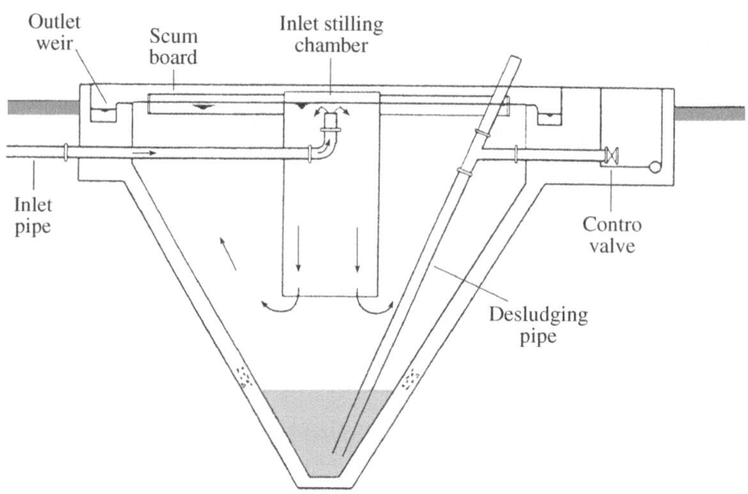

Fig. 3.5 – Vertical flow settling tank (after Gray, 1999)

Vertical flow tanks have a minimum cone angle of 60° to ensure that sludge does not accumulate on the walls, thus the resulting depth of the cone limits the size of these tanks to a maximum of about 4m diameter. For larger designs radial flow settling tanks are preferred (Figure 3.7), these include a centre driven rotating scraper (*skimmer blade*) to collect scum on the surface (Figure 3.8) and floor scraper to gather sludge into a central slump for desludging.

Ex 3.1
Which of the following are important parameters in the process description of sedimentation:
a) height of the tank
b) surface area
c) material of construction
d) flow rate
e) air temperature

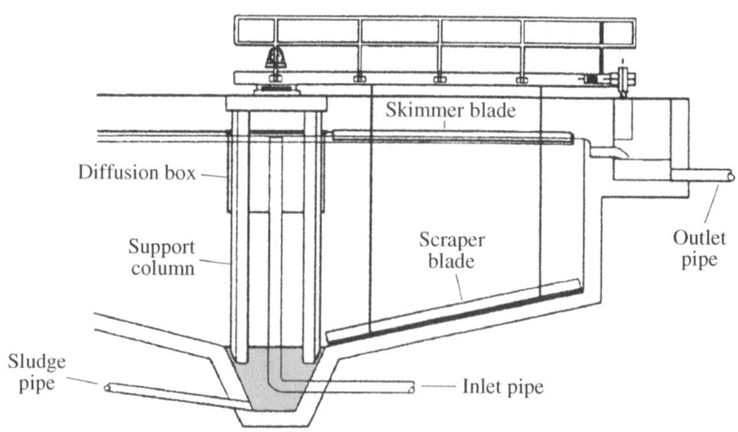

Fig. 3.6 – Radial flow settling tank (after Gray, 1999)

Fig. 3.7 – Skimmer blade on a radial flow settling tank

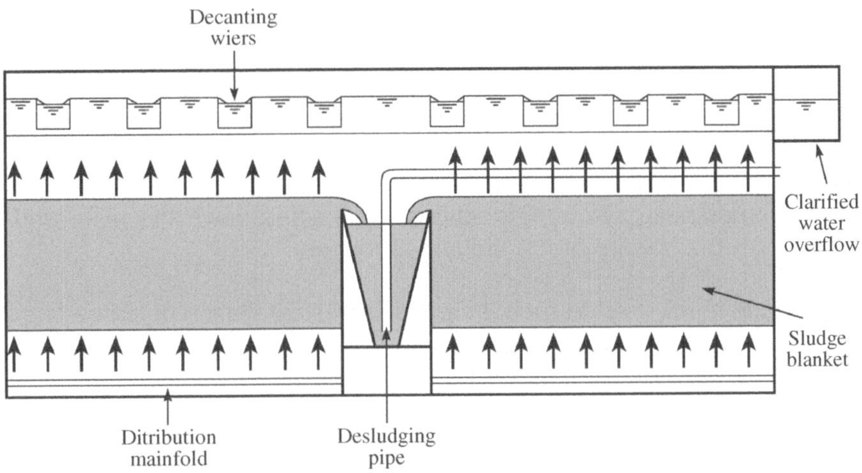

Fig. 3.8 – Flat-bottom sludge-blanket clarifier (after Gray, 1999)

3.2.3 Clarifiers

Clarifiers are employed in both water and wastewater treatment for sedimentation of flocculent particles formed during coagulation-flocculation (Unit 4) and removal of effluent biomass from biological wastewater treatment (Unit 6).

The formation of flocculent precipitates or *flocs* during coagulation-flocculation is an integral part of the overall solids removal process in water treatment. The contacting between newly formed flocs with old sludge assists in maximising the settling rates of the particles. This is achieved in circular or flat-bottom *sludge-blanket clarifiers* by suspending a *blanket* of sludge in a vertical flow clarifier (Figure 3.8). The water passes through the sludge-blanket and overflows into decanting weirs; the height of the sludge blanket is maintained by removing sludge periodically via desludging pipes.

Sludge-blanket clarifiers are also applied in the separation of activated sludge solids from mixed liquors (known as *secondary clarifiers*). The flocculant solids tend to form sludge blankets allowing the clarified effluent to overflow through surface weirs. Secondary clarifiers are typically circular in design (Figure 3.9) with scraper that collect the settled sludge to central sump for removal.

Fig. 3.9 – Secondary clarifier (courtesy of Sydney Water, Australia)

One of the crucial parameters in sedimentation is the surface area available for settlement (*Process Sciences and Engineering for Water and Wastewater Treatment* – Unit 3). This can be increased by the addition of *lamella plates*. These are flat plates (although some designs use corrugated plates or tubes of various cross sections) angled at 60° and fixed on a 50mm pitch (Figure 3.10). Water flows upwards between the plates and solids settle onto the plate immediately below. The sludge then slides down the plate and is removed from the bottom of the unit via a hopper.

The technique lends itself to *retrofitting* that is the installation of plates or tube modules into existing clarifiers to increase the throughput (Figure 3.11).

LAMELLA PLATES
Lamella plates do not increase the settling velocity, they simply increase the area available for settlement without increasing the footprint of the process.

Fig. 3.10 – Lamella plate (courtesy of Anglian Water, UK)

When the size of a sludge-blanket clarifier is a limitation due to the tank footprint, coagulation, flocculation and clarification can be accommodated in a single *sludge recirculation* clarifier (Figure 3.12). In this system the central zone provides mixing of influent coagulated water with old sludge and the outer section provides settlement and removal of sludge. The equipment is mechanically complex and, because of the recirculation and scraper drives,

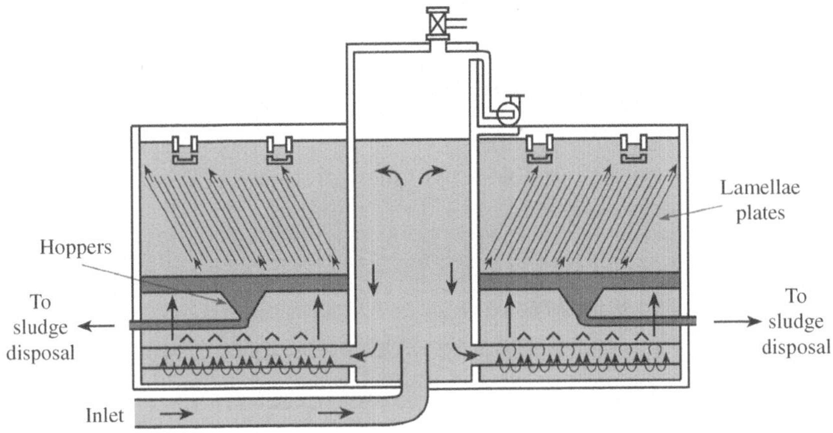

Fig. 3.11 – Lamella plate clarifier (courtesy of Anglian Water, UK)

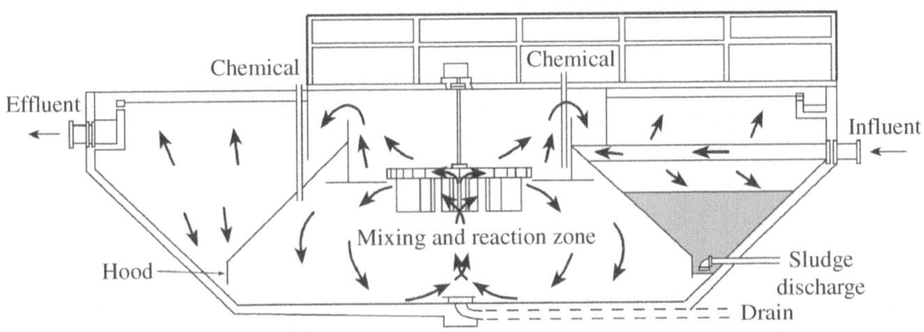

Fig. 3.12 – Solid recirculation clarifier (after Droste, 1997)

more expensive to operate than conventional vertical and radial flow units. Solids recirculation clarifiers are more common in industrial water treatment applications than in municipal water treatment works.

Sludge-blanket and sludge recirculation clarifiers are generally designed to handle flocs which are fairly consistent in size and density and, typically, at concentrations in the range 20–100 mg/l. A well designed and operated clarifier should produce clarified water with suspended solids around 5–10 mg/l.

Section 3.3 requires a study time of about 1 hour

3.3 Flotation

It is sometimes desirable to encourage certain suspended solid particles in water to float rather than to sink. *Dissolved air flotation* is a process widely used in water and wastewater treatment to achieve this and has benefits, by comparison with conventional sedimentation processes, in terms of footprint and waste sludge concentration.

Dissolved air flotation (DAF) is used for a variety of water and wastewater treatment processes including:

- clarification after flocculation of moorland surface waters for treatment to potable quality;
- treatment of industrial effluents;
- protein recovery from meat processing and dairy wastes;
- removal of oil from refinery effluents and
- thickening of sludge from sewage treatment processes (Figure 3.13).

Ex 3.2
Which of the following applications are not suitable for DAF:
a) removal of oil from refinery effluents
b) removal of fats, oils and greases
c) separation of gold from a mineral slurry
d) separation of protein from meat waste
e) thickening of sewage sludge

Fig. 3.13 – Dissolved air flotation thickener (courtesy of Sydney Water, Australia)

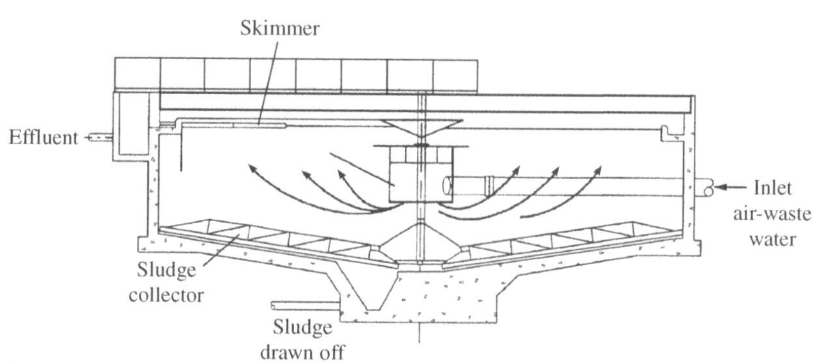

Fig. 3.14 – Dissolved air flotation (after Metcalf and Eddy, 1991)

DAF works by attaching air bubbles to flocs so as to reduce the density of the combined bubble floc agglomerate. When this is reduced to below that of water the floc rise to the surface where sludge accumulates or *floats* and can be removed by skimming off.

Small DAF units are generally circular (Figure 3.14) but larger units are generally rectangular.

Performance parameters for DAF are generally similar to those for sedimentation clarifiers but *surface overflow rates* or *rise rates* are typically five times those achieved in sedimentation, giving a smaller footprint. Sludge concentrations are around 2–4% compared with 0.5–2% for sedimentation clarifiers.

Section 3.4 requires a study time of about 1 hour

3.4 Screening

Screening is the removal of solids by passing the water through apertures, which are smaller than the solid particles (Figure 3.15). Screens are used for many applications in water and wastewater treatment including:

- removal of gross solids which might block downstream processes;
- removal of abrasive solids which might damage pumps and similar equipment;
- protection of potable water works intakes;
- removal of algae from potable water;
- "polishing" of sewage works final effluent;
- removal of floating material from industrial wastewaters and
- between unit operations in wastewater treatment.

SCREENS
Screening proves protection for downstream processes by removing large solid materials.

Fig. 3.15 – Screening of raw sewage (courtesy of Anglian Water, UK)

Screens are usually classified according to the size of the apertures as shown in Table 3.1.

Table 3.1 – Screen classification

Type	Aperture (mm)
Coarse	>50
Medium	15–50
Fine	3–15
Milli	0.25–3
Micro	0.025–0.2

Screens generate waste materials or *rags* (Figure 3.16) that are normally dewatered and transfer to a skip (Figure 3.17) for landfill disposal. Alternative strategies include disintegrating and redirecting to the main flow and biological treatment.

Fig. 3.16 – Screen wastes (courtesy of Anglian Water, UK)

Fig. 3.17 – Collection of screen wastes (courtesy of Yorkshire Water, UK)

3.4.1 Bar Screens

Bar screens, also known as *rack* or *coarse screens* normally consist of parallel bars with spaces. The bars can be rectangular or tapered. 12 mm is now typically considered a coarse opening for raw sewage. These screens can be cleaned either by hand or mechanically. Mechanical cleaning uses automatic rakes that enter the spaces and lift the screenings off for removal.

BAR SCREEN
Bar screen are mechanically cleaned by *rakes*.

3.4.2 Travelling and Rotating Screens

There are a number of proprietary *travelling screen* systems that rely upon the whole screen moving, an action that helps to prevent blinding and headloss build-up due to rags on the screens. Often these devices consist of a series of teeth, made of metal or nylon, travelling along a continuous belt (Figure 3.18).

Such systems are today often selected for coarse and medium duties as they are manufactured as standard units.

TRAVELLING SCREENS
Travelling screens are often selected for coarse and medium screening.

Fig. 3.18 – Continuous belt screen (courtesy of Thames Water, UK)

Fine screens for tertiary sewage treatment and *microstrainers* for potable water treatment are usually of the rotating type consisting of a low speed rotating cylindrical screen through which the water passes, which is usually continuously backwashed by jets of screened water.

Fig. 3.19 – Rotating drum screen (courtesy of Huber, Germany)

Rotating screens are called *cup screens* when the flow is from the centre of the screen outwards and *drum screens* when the flow is from the outside of the cylinder to the inside (Figure 3.19). The screen itself is often in the form of a wire mesh (Figure 3.20), but this material can be easily damaged, block readily and may distort under pressure.

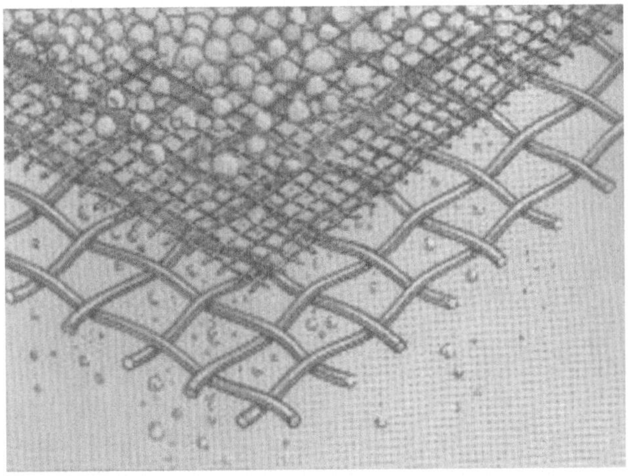

Fig. 3.20 – Wire mesh screen

An alternative screen material is *wedge wire* usually manufactured in fixed mesh sizes (Table 3.2) and made from stainless steel. Because the material is wedge shaped in cross section (Figure 3.21), the gap between the strands is wider on the downstream side of the screen than the minimum screen pitch or mesh size so that it is self clearing and less likely to block.

Table 3.2 – British standard sieve meshes

Sieve	μm	Sieve	μm	Sieve	μm
4	4000	12	1400	30	500
5	3350	14	1200	36	420
6	2800	16	1000	44	355
7	2400	18	850	52	300
8	2000	22	710	60	250
10	1680	25	600	100	150

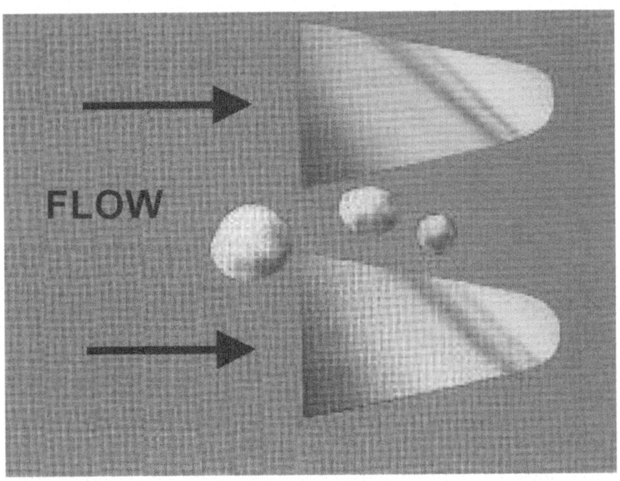

Fig. 3.21 – Wedge wire

Wedge wire is available in slot sizes down to 50 μm and is frequently formed into cylindrical screens, which fit directly onto the end of a pipe for submerged water intakes from rivers. In this application cleaning is usually by backflushing with a combination of water and compressed air.

3.5 Comminution

Section 3.5 requires a study time of about 1/4 hour

Comminution is the mechanical process by which the size of suspended solid particles is reduced. Comminution is used wherever the size of solid particles suspended in an aqueous stream needs to be reduced to improve the performance or to prevent clogging of downstream processes. The nature of the solids must be relatively soft to avoid wear on the comminutor blades but typical applications include:

- domestic sewage containing rags, paper and other organic debris and
- wastewater from the meat processing industry containing meat residues.

A typical comminutor configuration (Figure 3.22) consists of a chamber which houses a cylindrical slotted screen which is rotated by an electric motor. The rotating screen is fitted with cutting teeth at intervals which pass through slots in a fixed stator providing a cutting action. Solid material which is too large to pass through the screen is collected on the surface of the cylinder and, as the cylinder rotates, reaches the fixed stator blades and is cut into smaller pieces. The resulting macerated material is small enough to pass through the screen slots. The unit is self cleaning so should not clog and the suspended solids in the treated water stream are all smaller than the slot size of the screen and will typically pass through a 6 mm screen.

Fig. 3.22 – Comminutor (courtesy of Yeomans Chicago Corporation, USA)

COMMINUTORS
Comminutors uses a rotating slotted screen to macerated solids into fine particles that become small enough to pass through the screen slots.

The advantages of the process are that:

- no screenings disposal is required;
- no detailed design is needed;
- the process protects downstream pumps and
- faecal matter is not removed by the process and can thus be treated by downstream processes.

The disadvantages are that the process:

- requires power;
- can be damaged by large objects;
- causes rags to “ball-up” and
- causes screenings material to appear in the sludge.

3.6 Filtration

Section 3.6 requires a study time of about 1 hour

Filtration uses *granular media* to remove low levels of suspended solids from water. Filtration systems are classified as either slow or rapid and can be operated either by gravity or under pressure.

Filtration is mainly used in potable water treatment and for pre-treatment in such applications as:

- filtration of borehole water to remove iron and manganese;
- *direct filtration* of upland waters after in-line coagulation;
- removal of residual suspended solids after clarification ;
- pre-filtration to protect membrane processes;
- removal of residual precipitates from industrial effluents and
- tertiary treatment of domestic wastewater.

3.6.1 Gravity Filtration

FILTRATION
Filtration systems remove low turbidity solids either by slow or rapid filtration using granular media such as sand.

Gravity filtration (solids removal from water) is a *batch process* that is carried out for a period typically around 24 hours, and the filter is then taken off line and cleaned by *backwashing*.

The usual filter medium is sand but other materials – anthracite, garnet, manganese oxide, dolomite and others – are used for special applications.

Slow sand filters have been used since Roman times for the purification of drinking water. Raw water is filtered through a bed of fine (0.25 mm) sand about 1 m deep. The flow velocity is slow (around 0.1 mh^{-1}) and on the surface of the sand a gelatinous layer form that removes turbidity, colour, taste and odours, by a combination of filtration and biological activity (bacteria, protozoan and algae), called the *schmützdecke*. As the schmützdecke builds up so the flow through the filter declines and, after a few weeks, the top layer of sand is removed by manual or automatic scraping. The sand is washed and subsequently returned to the filter. The quality of treated water is excellent but, because of the high land area required, the capital cost is very high and the process is little used in developed countries.

Rapid filtration was developed to provide some of the benefits of slow sand filtration but using a much smaller area. Typical approach velocities are in the range 5–20 mh^{-1} using 0.8 mm sand. The sand grains provide a filter with an effective screen size of about $^1/_{14}$th the diameter of the sand grains that is about 50 μm for normal 0.8 mm sand. However, the sand bed removes particles much smaller than this including *cryptosporidium* oocycsts (5 μm) and some bacteria (0.5 μm) by a process called *depth filtration*. As water flows downwards between the sand grains, particles larger than 50 μm are trapped by simple straining or *surface filtration*, which is the same process as is used in screening. Smaller particles are deposited on the surface of sand grains by a variety of processes including: direct interception, where particles are carried on to the grains by the water flow stream lines; diffusion, where random movements (*Brownian motion*) of the particles across the stream lines cause collisions between particles and sand grains; inertial deposition, where the particle's inertia carries it into collision with the sand grain when the stream line changes direction; sedimentation, where the particle's mass causes it to settle from the stream line onto the sand surface; and hydrodynamic forces such as swirl and turbulence (Figure 3.23).

Ex 3.3
How many times faster is the flow velocity in a rapid gravity filter compared to a slow sand filter:
a) The same
b) 10 times
c) 1000 times
d) 100 times
e) 10000 times

DEPTH FILTRATION
Depth filtration is a filtration process whereby the entire medium is involved in the filtration so that the quality of water improves with depth.

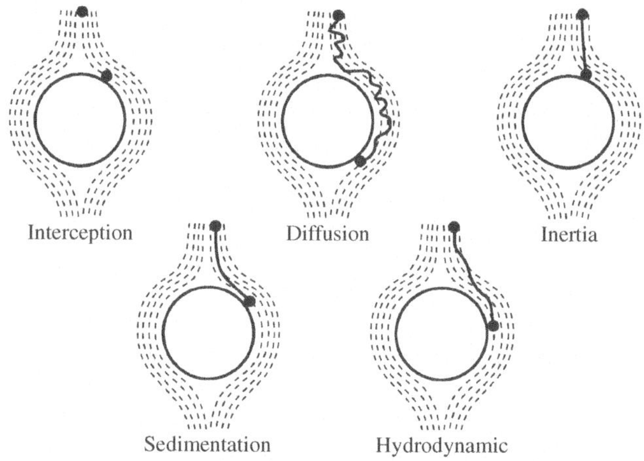

Fig. 3.23 – Depth filtration mechanisms (after Ives, 1982)

Once the particle has been transported to the surface of the sand grain it is held there by weak Van der Waal's forces. As the filter *run* proceeds the removed solids take up more and more of the available deposition sites on the sand grains and the *penetration* of solids moves further down the bed. Eventually the *capacity* of the filter is reached and suspended solids begin to appear in the filtered water and are detected as *turbidity breakthrough*.

RAPID GRAVITY FILTERS
Rapid gravity filters use coarse sand to filter suspended solids via depth filtration.

The sand bed is contained in a *filter shell* which may be circular or rectangular in plan and may be open topped operating under gravity flow (*gravity filter*) as shown in Figure 3.24 or closed and operated under pressure (*pressure filter*) as shown in Figure 3.25.

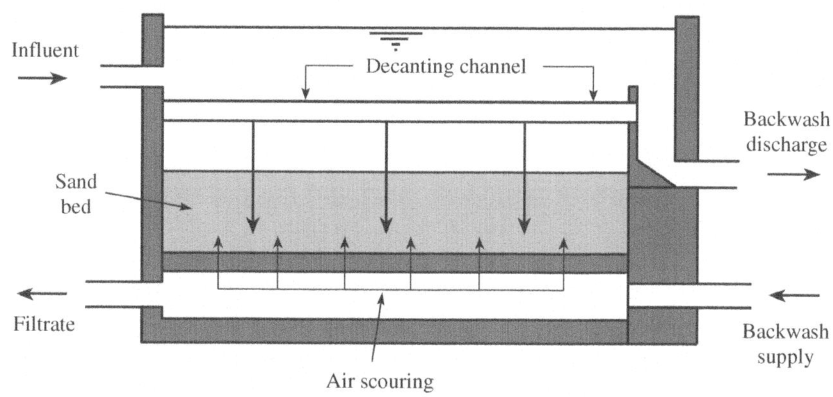

Fig. 3.24 – Rapid gravity filter (after Gray, 1999)

Ex 3.4
What parameter is used to determine the operating capacity of rapid gravity or pressure filters?

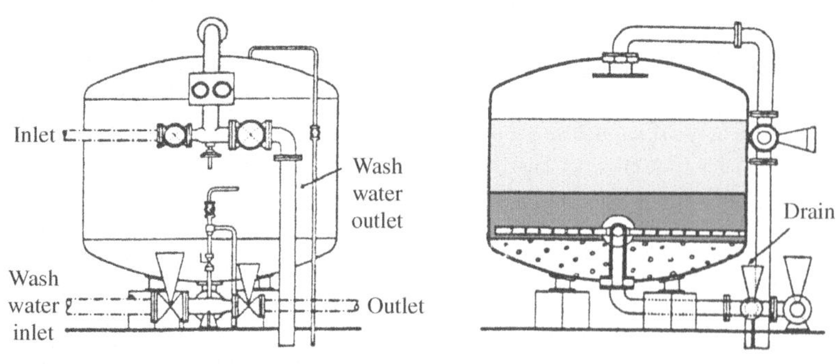

Fig. 3.25 – Pressure filter (after Gray, 1999)

Rapid sand filter runs are typically around 24 hours and backwashing takes about ½–¾ hr.

As solids accumulate in the filter bed the water flow through the bed is restricted causing an increase in the *head loss* (*Process Sciences and Engineering for Water and Wastewater Treatment* – Unit 11). In a gravity filter this head is provided by the static head of water above the sand bed whilst in a pressure filter it is pump head.

When the head loss becomes too high, or when breakthrough is detected, the filter run is terminated. The filter is then taken off line and *backwashed*. Backwashing is carried out by pumping filtered water upwards through the sand bed. When the upflow velocity of the wash water reaches the terminal velocity of the sand grains, the bed starts to *fluidise*, that is the sand grains begin to separate from each other and to float freely. The fluidisation of the sand releases the retained suspended solids, a process assisted by bubbling air through the sand bed (*air scouring*) which causes the sand grains to rub against each other dislodging the retained solids from the sand surface (Figure 3.26). The suspended solids are then washed from the sand bed and out of the shell to waste. Backwashing typically takes about 30–45 minutes.

BREAKTHROUGH
When suspended solids appear in the filtered water resulting from the filter reaching its maximum capacity.

AIR SCOURING
Air scouring is used to dislodge the retained suspended solids from the sand surface.

Fig. 3.26 – Air scouring of a rapid gravity filter (courtesy of Thames Water, UK)

Fig. 3.27 – Battery of pressure filters (courtesy of Anglian Water, UK)

Because filtration is a batch process, filters are usually built in *batteries* consisting of several filters (Figure 3.27) so that filtration to service can continue while one filter is being backwashed.

3.6.2 Continuous Filtration

An alternative to this batch filtration is *continuous sand filtration* (Figure 3.28). Here water enters the filter at the bottom of the sand bed and flows upwards. Filtration occurs in the upflow mode and filtered water is discharged from the top of the filter. Sand is continuously removed from the bottom of the bed and transported by an air lift pump to the top of the unit where it is washed with filtered water. The cleaned sand is returned to the top of the bed and the dirty washwater is discharged to drain. The amount of washwater used is the same as in a conventional batch filter but is discharged continuously at a low flow rather than as a large volume in a short time. The continuous filter is more expensive but no *standby* unit is required and there is no stoppage time for backwashing.

Fig. 3.28 – Continuous sand filter (courtesy of WesTech Engineering, USA)

3.6.3 Multi-media Filtration

DUAL MEDIA FILTRATION
Dual media filters allow for progressive filtration due to different particle sizes.

One method of optimising filter performance is to use two different filter media in the same filter bed (*dual media filtration*). A layer of coarse anthracite on top of a layer of fine sand will provide progressively finer filtration and, because anthracite has a lower density than sand, the coarse layer will remain on top after backwashing. In critical industrial water filtration applications, *multi-media filters* are used which have three media layers: coarse, low density anthracite at the top; conventional sand in the middle; and very fine high density garnet at the bottom.

Influent suspended solids to deep bed filters should not exceed 20 mgl^{-1} and preferably should be less than 10 mgl^{-1}.

Using 16/30 mesh sand (0.5–1.0 mm) deep bed filters should remove $>98\%$ of all particles larger than 2 μm but will usually achieve better performance than this. Typical filtered water suspended solids will be less than 2 mg/l and turbidity will normally be less than 1 NTU.

Section 3.7 requires a study time of about 1/2 hour

3.7 Centrifugal Separation

Centrifugal separation uses imparted angular velocity to create a high acceleration to increase the rate at which suspended solid particles are removed from water.

Angular velocity can be created in two ways:

- in the *centrifuge* it is produced by rotating a cylinder at high speed and
- in the *hydrocyclone* the cylinder remains stationery but the influent water enters tangentially at high velocity creating a swirling motion.

Whichever technique is used there is a significant energy input – in the case of the centrifuge it is the power for the motor which drives the cylinder whilst in the hydrocyclone it is the pumping power needed to generate the high inlet velocity.

Centrifuges are widely used in water and wastewater treatment for sludge dewatering (Unit 8).

Hydrocyclones are used for solids removal in water and wastewater treatment. Two areas where the process has gained acceptance is in the removal of solids in stormwater overflow systems and grit removal in wastewater treatment.

Hydrocyclones for grit removal are known as *vortex separators* and *Pista traps.* Sewage enters a circular tank tangentially to create a vortex and grit falls to bottom with the effluent outlet at the centre (Figure 3.29). Pista traps incorporate power driven paddles with air lift to remove grit. The advantages are that clean grit is produced and only a small area is required.

Ex 3.5
How do hydrocyclones differ from centrifuges in term of separation technique?

Fig. 3.29 – Circular grit trap (courtesy of Huber, UK)

A similar technique is used in the *aerated grit chamber*. In this process an air flow encourages a helical flow pattern (*roll*) in a tank (Figure 3.30). Not only does this give a smaller footprint than a constant velocity channel but also aerates the sewage. This process is very efficient at grit removal and produces a very clean grit but does require an air supply, which adds to the cost.

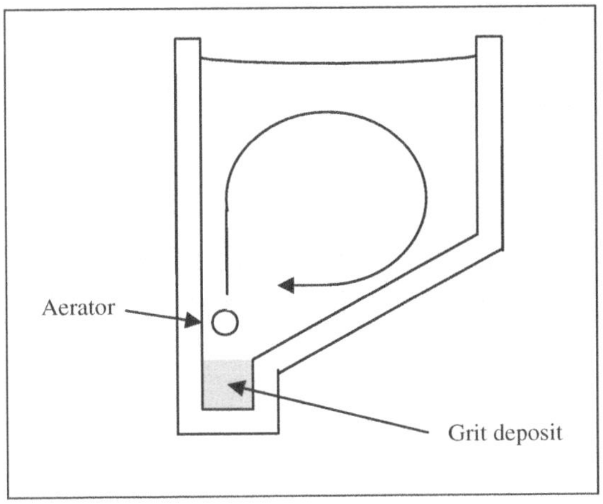

Fig. 3.30 – Aerated grit chamber

3.8 Self Assessment Questions

SAQ 3.1 Describe the principal features of a lamella clarifier and what are the advantages of lamella clarifiers?

SAQ 3.2 Give three examples of depth filtration mechanisms.

How are deep bed filters cleaned?

What are the advantages and disadvantages of continuous sand filters?

SAQ 3.3 For the following processes explain the mechanisms behind the process:

a) Dissolved air flotation
b) Hydrocyclone
c) Sludge blanket clarifier
d) Wire mesh screen

SAQ 3.4 Describe the principal features of a lamella clarifier and what are the advantages of lamella clarifiers?

SAQ 3.5 Give three examples of depth filtration mechanisms.

How are rapid gravity filters cleaned?

What are the advantages and disadvantages of continuous sand filters?

3.9 Solutions to Exercises

Ex 3.1 b, c

Ex 3.2 c, this is coarse bubble flotation.

Ex 3.3 d

Ex 3.4 Suspended solids in the filtered water from a gravity or pressure filter is used to determine the filtration capacity by measuring for turbidity breakthrough.

Ex 3.5 Both techniques use angular velocity to separate solids. In hydrocyclones the cylinder remains stationary whereas in centrifuges the cylinder is rotated at high speeds in order to generate the angular velocity.

Bibliography

Unit 3 – Physical Processes

Droste R.L. *Theory and Practice of Water and Wastewater Treatment.* John Wiley & Sons Inc., Chichester (1997).

Metcalf and Eddy, *Wastewater Engineering, Treatment and Reuse.* 4th ed., McGraw Hill, New York (2003).

Sincero A.P. and Sincero G.A. *Physical-chemical Treatment of Water and Wastewater.* IWA Publishing, London (2002).

4. Chemical Processes

Contents

Aims and Objectives

This Unit covers the principles of chemical processes used in water and wastewater treatment:

After studying these notes you should be able to:

1. explain in your own words the following terms and concepts:

   pH adjustment
   precipitation
   coagulation
   flocculation
   oxidation/reduction
   disinfection

2. describe the different unit processes used in the treatment of water and wastewater.

It is important that you are able to complete all the self assessment questions at the end of this Unit.

Essential Prerequisites

It is not necessary to have a completed any other Units before undertaking this Unit.

Additional, information on the fundamentals for this Unit, refer to the following Units in *Process Science and Engineering for Water and Wastewater Treatment*:

Unit 1 Fundamentals of Water Chemistry
Unit 2 Chemical Kinetics and Equilibria
Unit 3 Colloids and Surface Chemistry
Unit 4 Fundamentals of Microbiology
Unit 7 Fundamentals of Process Engineering

4.1 Introduction

By changing the chemistry of water and wastewater, the altered properties of the water can be used to provide the basis for the operation of different water and wastewater unit processes.

The dosing of specific chemicals is used in order to enhance precipitation during sedimentation and different chemicals are involved in the disinfection of water and wastewater in order to control the number of bacteria and viruses prior to discharge to the environment or distribution. Chemicals are also used to inhibit the formation of specific compounds and improve the dewatering of sludge.

In this Unit the principle chemical processes involved in solids removal and the disinfection of water and wastewater are outlined.

Section 4.2 requires a study time of about 1 hour

4.2 Chemical Solution Dosing

When chemicals need to be added to a process stream they are usually added in the form of a solution. The *bulk chemical* may be purchased as a liquid or in the solid form as a powder or crystal and stored on site in bulk storage tanks. Solids are mixed to form a solution either manually or automatically in *mixing tanks*. It is usual to transfer solutions from mixing or bulk storage tanks to day tanks for dosing.

The purpose of a *chemical dosing system* (Figure 4.1) is accurately to meter the necessary amount of chemical to be added and to ensure that it mixes rapidly and thoroughly with the process stream.

Fig. 4.1 – Dosing pumps and storage tanks (courtesy of Anglian Water, UK)

The chemical is usually added as a solution using a metering pump. This is a *positive displacement* pump using either a *piston* or *diaphragm* head to deliver an accurately repeatable volume of solution on each pump *stroke*. The speed and/or the length of the delivery stroke are adjustable to vary the volume delivered. This adjustment may be manual or automatic.

Having added the controlled *dose* it is necessary to mix it into the process water flow so that the reaction can proceed. Since chemical doses are usually very small volumes by comparison with the process water flow, this mixing is very important if the reaction is to be completed quickly. The reactions between acids and alkalis are generally rapid, although there are some exceptions like the use of lime as an alkali which, because it is relatively insoluble, is usually dosed as a slurry rather than a solution.

When reactions are sufficiently rapid we can mix the dosed chemical by using a *static* or *hydraulic mixer*. In static mixers, blades are placed into the pipe, which induce turbulence through the dissipation of energy. These normally provide up to 98% complete mixing within three pipe diameters downstream. In hydraulic mixers, jumps are placed into an open channel (Figure 4.2) and mixing is accomplished as a result of turbulence that exists in the flow regime.

Fig. 4.2 – Hydraulic mixer (courtesy of Sydney Water, Australia)

Alternatively, chemical mixing can be provided by continuous rapid mixing using a *flash mixing tank*. This is simply a tank, ideally of circular cross section, providing a residence time of about 1–2 minutes and fitted with some sort of mechanical mixer, typically a propeller, turbine or paddle, which provides a high shear gradient of $>1000s^{-1}$.

Ex 4.1
What forms of rapid mixing are employed during chemical dosing?

Once the chemicals have been dissipated throughout the water, slow mixing is used to increase the chances of collision between floc particles known as flocculation (Section 4.4). Figure 4.3 shows a typical paddle type flocculator.

Fig. 4.3 – Paddle mixer (courtesy of Sydney Water, Australia)

In order to achieve a consistent performance it is almost always necessary to provide some form of automatic control of chemical dosing. Such control systems are loop systems in which there is a *desired value* of a variable parameter, say pH, a *measured value* of the same variable and some sort of mechanism which compares the two and adjusts the chemical dosing to correct any difference.

There are two types of control loop: in the *open loop*, or *feed forward* or *predictive* control system, the measurement and comparison are carried out upstream of the corrective action whilst in the *closed loop* or *feed back* control system the measurement and comparison are carried out downstream of the corrective action.

Open loop systems (Figure 4.4) are inherently unstable because there is no means of confirming that the corrective action has had the desired effect. They are, however, useful in the case of, for example, flow proportional control where the rate of dosing is increased or decreased as the flow to the process changes.

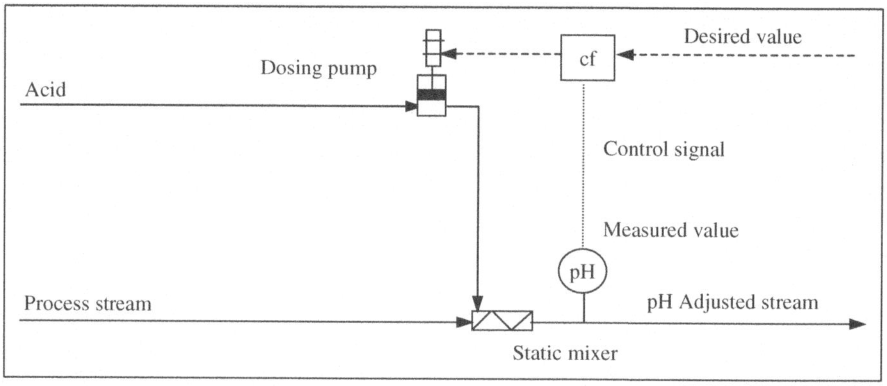

Fig. 4.4 – Open loop control

Open loop control requires that we know how a controlled change to the dosing rate will affect the measured variable. For example, in a pH adjustment system, if the desired value of the pH is 7 and the measured value is 5 it is necessary to know how much alkali to dose to restore the pH to 7. This is usually a totally unpredictable relationship and, for control of pH, closed loop control (Figure 4.5) is almost universal.

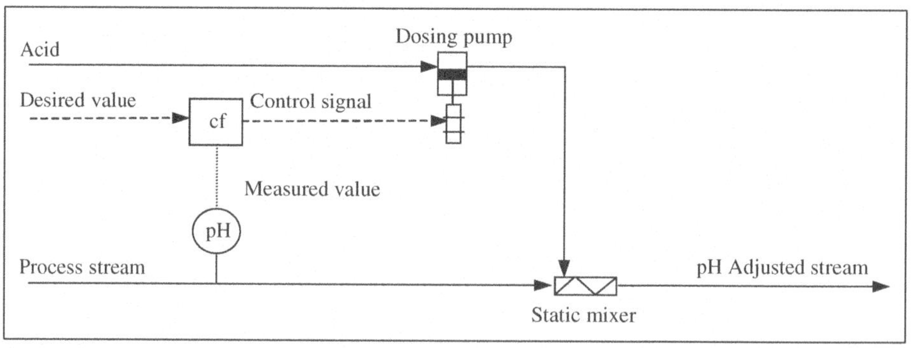

Fig. 4.5 – Closed loop control

SINGLE SET POINT
Single point control chemical dosing system switch on and off around a desired value.

The simplest form of control is a *single set point* which starts dosing if the set point (desired value) is exceeded and stops dosing when the measured variable falls below the desired value. In all but the most consistent waste streams this form of control will lead to rapid switching on and off of the dosing system (*hunting*) and this causes both instability and unnecessary wear on the equipment.

In-band or *two position* controls introduce a hysteresis into the control system by operating between two set pH control points either side of the desired value. If an acidic stream is being dosed with alkali then the alkali addition starts if the pH falls below the lower set point and stops if the pH rises above the upper set point. Careful design of the dosing system should ensure that for most of the time the control will be in the *dead band* between the two set points. The width of the dead band will be determined by the allowable limits on the pH, bearing in mind that when dosing is started or stopped there will, inevitably, be some *over-shoot*.

Under *proportional* control any deviation from the desired value produces an output which is linearly proportional to the magnitude of the deviation and which can be used to *modulate* the chemical dose (i.e. the greater the deviation the greater the corrective action).

The response of a control system can be improved by the addition of two other types of action called integral and derivative control. The resulting system is called three term control or PID (*Proportional plus Integral plus Derivative*).

Integral action provides a persistent signal after the error has been eliminated. This signal is a function of the error and the time for which the error persists and corrects the offset. However, it increases the maximum deviation and lengthens the period of oscillation given by proportional control alone. Derivative or anticipatory action reduces the oscillation produced by proportional control alone but does nothing to correct the offset. Combining the three terms in the correct proportions (tuning the loop) optimises the response to minimise the offset, peak deviation and length of oscillation.

Control action is not immediate. The time taken for a control loop to respond to a change depends on:

- the *measurement lag*, which is the time the measuring instrument takes to detect a change in measured variable;
- the *control lag*, the time it takes for the control element (modulating valve or dosing pump) to reach its new position after receiving a new control signal and
- the *process lag*, the time it takes for a change in dosing to create a change in the pH which includes the hydraulic retention time between the dosing point and the measuring point, the reaction time and the mixing time.

The loop time effectively represents the period of time for which no control is applied and the longer the loop time the longer the greater the overshoot and the higher the amplitude of the oscillations occurring after a disturbance.

Whilst it is quite usual to control on only one variable, such as pH, sometimes the response of a control system can be improved by controlling against two variables. The most common example is where a water or waste stream has a fairly consistent composition but where there are significant changes in flow. Consider an acid stream being neutralised by lime addition with a total process time of 15 minutes. An increase in stream flow will result in a decrease in pH because the lime dose is insufficient. The pH monitor will detect this change and respond to it by increasing the dose but it will be 15 minutes before this corrective action is commenced and much longer for the pH to be restored to the desired value. If a feed forward signal from a flow meter is provided the dose rate can be increased proportionally to the flow rate before any change in pH is detected. This will significantly reduce the loop time.

MULTI-VARIABLE CONTROL
Multi-variable control permits the control of a number of variables during chemical dosing such as mixing and dosage concentration.

This type of multi-variable control can be applied in a number of ways. It is quite common in the case of dosing pumps to control the motor speed in proportion to the flow and the pump stroke in proportion to pH. In the case of on/off valves the frequency of operation can be controlled in proportion to flow and the duration of opening in proportion to pH. Alternatively the two signals can be summed with "weighting" factors so that, for example, the control signal is 80% of the feed forward flow signal and 20% of the feed back pH signal.

A well designed dosing system should control the chemical dose to ±2% and achieve at least 98% mixing.

4.3 pH Adjustment

Section 4.3 requires a study time of about 1 hour

It is often necessary to adjust the pH of a water or wastewater. This may be to optimise the chemical conditions for a reaction or avoid corrosion or scale deposition. The pH of a solution pH can be changed with the addition of acids or alkalis or the use of bicarbonate as a "buffer" (*Process Science and Engineering for Water and Wastewater Treatment* – Unit 2).

There are many reactions in water and wastewater treatment which require the pH to be adjusted to a particular value or, more correctly, within a band. Typical examples include:

- precipitation of heavy metal hydroxides;
- coagulation of colloidal material;
- prevention of carbonate scale deposition in reverse osmosis;
- neutralisation of wastewater to meet consent limits;
- to avoid corrosion in cooling water systems and
- to control the formation of odours in sewer systems.

The most commonly used reagents for pH adjustment are sulphuric and hydrochloric acids, calcium hydroxide (lime), sodium hydroxide (caustic soda) and sodium carbonate (soda ash).

The performance of a pH adjustment system will depend on:

- the chemical reactions involved;
- the accuracy of the metering pump;
- the efficiency of the mixing device;
- the precision of the pH measuring element and
- the type of automatic control system employed.

Typically most industrial systems are capable of controlling to ± 0.1 pH unit. An example is shown below for lime dosing where the final water pH is dependent on lime dosed and mixing time (Figure 4.6)

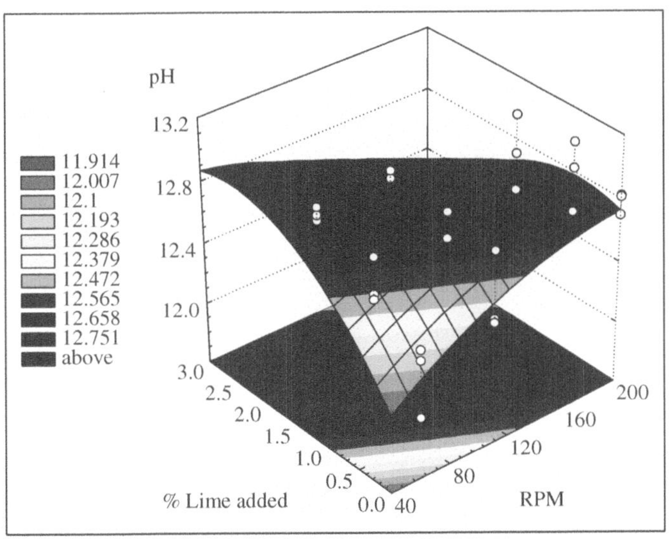

Fig. 4.6 – Conditions affecting the control of pH when dosing lime

Section 4.4 requires a study time of about 1/4 hour

4.4 Precipitation

The addition of appropriate cations or anions to a water or wastewater can be used to change the chemistry of that solution due to creation of insoluble salts. This process is referred to generically as *precipitation* (*Process Science and Engineering for Water and Wastewater Treatment* – Unit 2). The insoluble salts can then be separated by physical processes (Unit 3).

PRECIPTATION
The addition of appropriate cations or anions to a water or wastewater can be used to create insoluble salts.

The most commonly used precipitation techniques are:

- addition of hydroxide or sulphide to remove heavy metal cations from wastewater: $Zn^{2+} + 2OH^- \Leftrightarrow Zn(OH)_2$
- partial softening of water by removing calcium bicarbonate (temporary hardness) by adding lime to precipitate calcium carbonate (*Clarke's process*): $Ca(HCO_3)_2 + Ca(OH)_2 \Rightarrow 2CaCO_3 + 2H_2O$
- total softening of water by the addition of lime and soda ash (*Lime-Soda Process*): $CaSO_4 + Na_2CO_3 \Rightarrow CaCO_3 + Na_2SO_4$
- removal of phosphorus from wastewaters by addition of ferric chloride: $Na_3(PO_4) + FeCl_3 \Rightarrow FePO_4 + 3NaCl$.

The most widely used precipitation reagents are lime (calcium hydroxide), caustic soda (sodium hydroxide), magnesia (magnesium hydroxide), sodium sulphide and soda ash (sodium carbonate).

The chemical is normally dosed by a metering pump (Section 4.2). The main exception is lime which, being sparingly soluble, is dosed as a slurry of about 2–5% by weight. It is essential to keep this slurry moving otherwise it settles out and blocks pumps and pipework. Progressive cavity pumps are widely used for lime slurry handling and dosing is often controlled by an automatic valve on a continuously circulating slurry ring main.

The pH of the solution is important because not all metal hydroxides show the same relationship between pH and solubility, as Figure 4.7 demonstrates. For automation of dosing alkalis for metal hydroxide precipitation, pH is an appropriate measured variable and Redox potential can be used for automation of sulphide dosing.

The amount of chemical required to be dosed will depend on the chemistry. Most reagents react stoichiometrically but lime, because it is dosed as a slurry and has poor kinetics, usually requires an excess of at least 20% to ensure that reactions are complete.

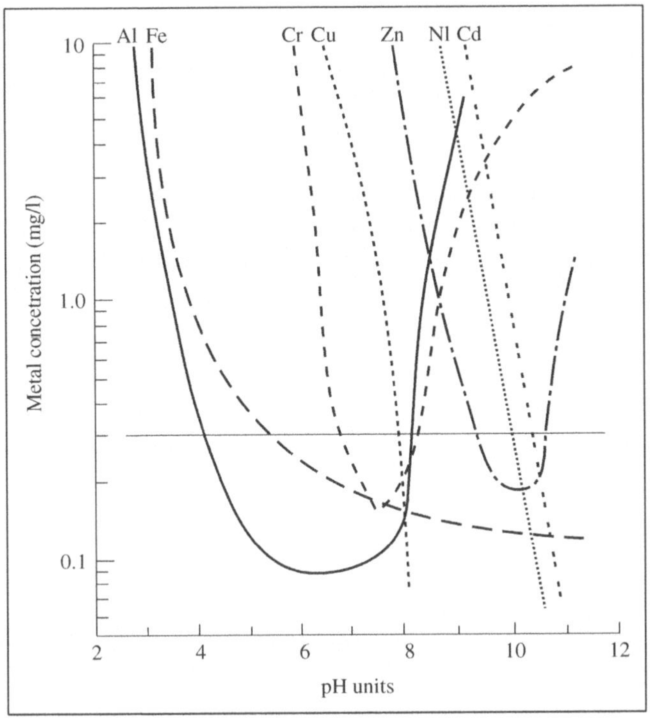

Fig. 4.7 – pH v solubility for metal hydroxides

Ex 4.2
Why is pH important in the precipitation of metals?

Section 4.5 requires a study time of about 1 hour

4.5 Coagulation and Flocculation

COAGULANTS
Coagulants are inorganic salts added to water or wastewater to neutralise the electrostatic charges on particles and initiate coagulation.

Coagulation is the first stage of the process of removing colloids, which cause turbidity and consists of neutralising the surface charge so that the particles can grow to a settleable size (Figure 4.8). Coagulation also assists in the removal of organic contaminants, particularly those which cause colour, taste and odour as well as bacteria and algae (*Process Science and Engineering for Water and Wastewater Treatment* – Unit 2).

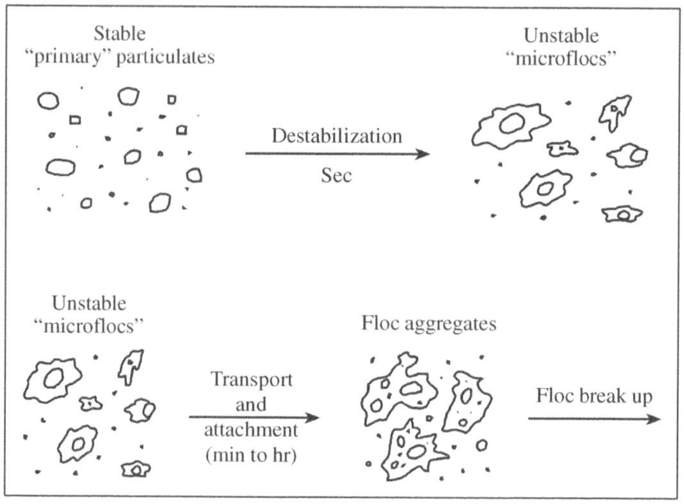

Fig. 4.8 – Floc formation and growth

Flocculation is the growth of the microscopic *flocs* formed by the coagulation of colloids into large particles which can be removed from the water. The precipitation of the coagulant as hydroxide floc is an essential part of this process but, in order for flocs to grow, they must be brought together so that they can collide and adhere to each other. Natural Brownian motion will achieve this (perikinetic flocculation) given time but, for treatment plant this needs to be speeded up either by the addition of chemical flocculants or by mechanical flocculation. Floc is an open structure containing large quantities of water which means that its density is only slightly higher than that of water. However, it undergoes a structural change with time, called syneresis, which results in it losing water and becomes more dense as it ages. Figure 4.9 shows representative water and wastewater flocs generated during coagulation.

There are many reactions in water and wastewater treatment which require coagulation and flocculation. Typical examples include:

- removal of colloidal material;
- removal of colour and natural organic matter;
- removal of emulsified oil in wastewater treatment;
- recovery of proteins from food industry wastes and
- thickening of sludges from sedimentation processes.

The most commonly used reagents for coagulation are salts of aluminium and iron and, occasionally, lime.

Since most colloids in waters and wastewaters are negatively charged it is necessary to add positive charges to the water to neutralise this surface charge. The concentration of a given electrolyte needed to effect coagulation is a function of the ionic charge (Schulz-Hardy rule). Thus divalent ions are more effective coagulants than monovalent and trivalent ones are more effective than divalent. The most common sources of trivalent positive ions are aluminium and iron (III), salts. Aluminium sulphate, iron chloride and iron sulphate are all readily available and iron (II) sulphate oxidised on-site by the addition of chlorine (*chlorinated copperas*) is also used.

FLOCCULATION
Flocculation is the growth of microscopic *flocs* formed by the coagulation of colloids into large particles.

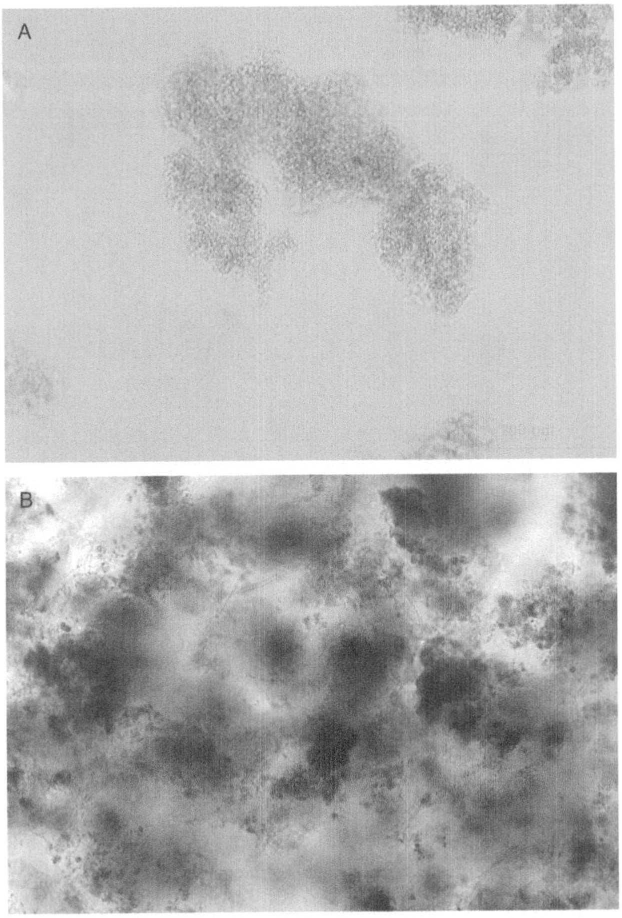

Fig. 4.9 – Flocs formed during the coagulation of water (A) and wastewater (B)

Adding a coagulant to destabilise a colloid is only the first step. The coagulant itself is an undesirable contaminant in the water and, ideally, it should be removed by precipitation as the hydroxide. Aluminium hydroxide is amphoteric and has a minimum solubility in the pH range 5.5–6.5, so careful control of the pH is critical when using aluminium salts as primary coagulants. The solubility of iron hydroxide falls with increasing pH but simply raising the pH does not necessarily give the best results. This is because the process is much more complicated than just destabilisation of the colloid and precipitation of the coagulant. The precipitated material or "floc" is not a simple hydroxide and organic material is often adsorbed onto the floc. As a result the best chemical conditions for coagulation vary widely from one water to another. Figure 4.10 shows an example for the same water coagulated with iron chloride at two different pH's.

Coagulant and pH adjustment chemicals are usually added, by means of a metering pump, at the same point in the system and must be mixed thoroughly and quickly. This is usually achieved by a *flash mixer* or by an in-line static mixer (Section 4.2). A typical system control is shown schematically in Figure 4.11.

Fig. 4.10 – Coagulation of a reservoir water rich with humic material at pH 4.2 (left) and pH 3.3 (right)

Ex 4.3
What application does jar testing serve in coagulation?

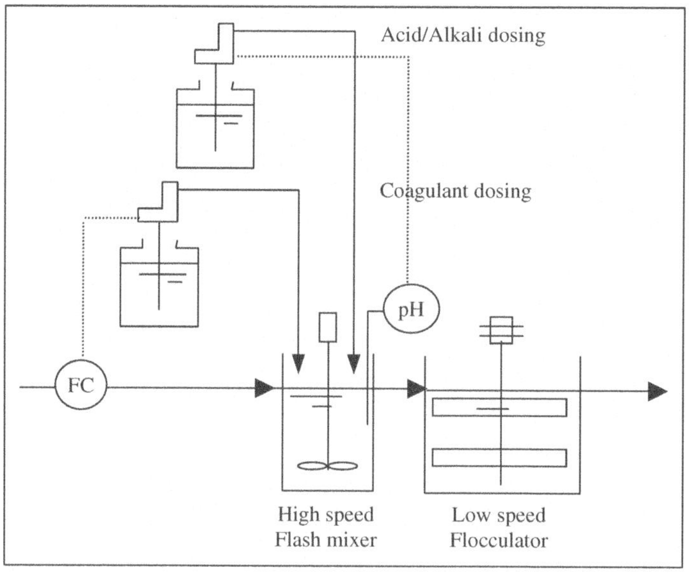

Fig. 4.11 – Coagulant dosing system

JAR TESTS
Jar tests are used to determine optimum coagulant concentrations and pH conditions for coagulation-flocculation.

Because water quality varies from day to day, therefore the optimum coagulation conditions vary as well. *Jar tests* (Figure 4.12) demonstrated the variation in floc formation due to varying coagulant dosing. These systems are typically carried out to optimise and establish operating conditions, but day to day coagulant and pH control is more commonly maintained by automatic control. Although systems based on zeta potential measurements are available, the simplest method of control is to provide coagulant dosing in proportion to flow only, with automatic pH control for acid or alkali dosing.

Mechanical flocculation uses a shearing mechanism to increase the chances of collision between floc particles (orthokinetic flocculation). This may be achieved by passing the coagulated water along a tortuous path – a serpentine pipe or baffled tank – or by the use of flocculation tanks with slow speed paddle stirrers (Figure 4.3). The power input into mechanical flocculators is quite critical. Too little agitation will not create enough collisions for the floc to grow whilst too much will break up the floc particles which do grow. The addition of some long chain chemicals (*polyelectrolytes*) will help the process by, effectively, acting as bridges between the micro-flocs as illustrated in Figure 4.13.

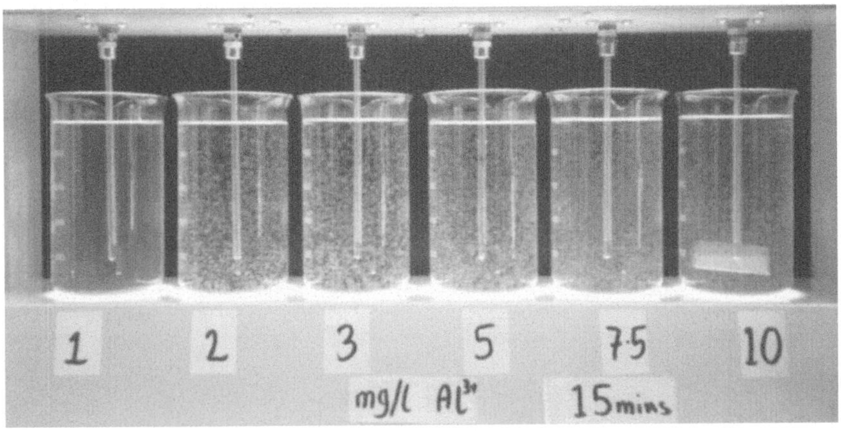

Fig. 4.12 – Coagulation flocculation at varying aluminium (Al) doses

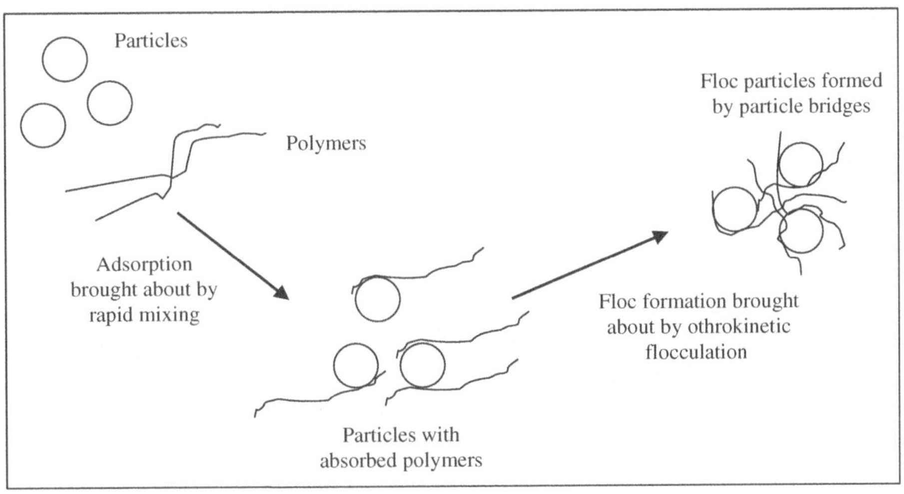

Fig. 4.13 – Mechanism of polymer bridging

Some natural chemicals, such as alginates, silicates and starches work to some extent but the most effective chemical flocculants are proprietary synthetic polymeric materials. Whilst the latter can be used in industrial water treatment, many of them are toxic and cannot be used for potable water treatment.

POLYELECTROLYTES
Polymers such as alginate are added during flocculation to aid in the bridges between the micro-flocs.

Polyelectrolytes are categorised by the charge carried by the active groups – cationic, anionic or non-ionic. Cationic polymers work well on negatively charged colloids and flocs whilst anionic polymers, similarly, work well on positively charged particles. Non-ionic polymers perform better than charged polymers on uncharged particles and sludges.

Polyelectrolytes are difficult materials to handle. Those that are supplied in liquid form are viscous and difficult to pump. Those supplied as powders are difficult to wet and form viscous non-Newtonian solutions. Special dosing systems are available for handling powdered materials.

Typical dose rates for primary coagulants in potable water treatment are about 1–5 mgl^{-1} as metal. Clay colloids tend to be removed best at pH levels around 6.0–7.0 but colour removal is better at 5.0–6.0. The most effective coagulant, the most effective dose and the optimum pH must be determined by jar testing. Higher dose rates, often as high as 50 mgl^{-1} as metal, are generally required for waste water treatment applications such as solids, organic and/or nutrient removal (Figure 4.14).

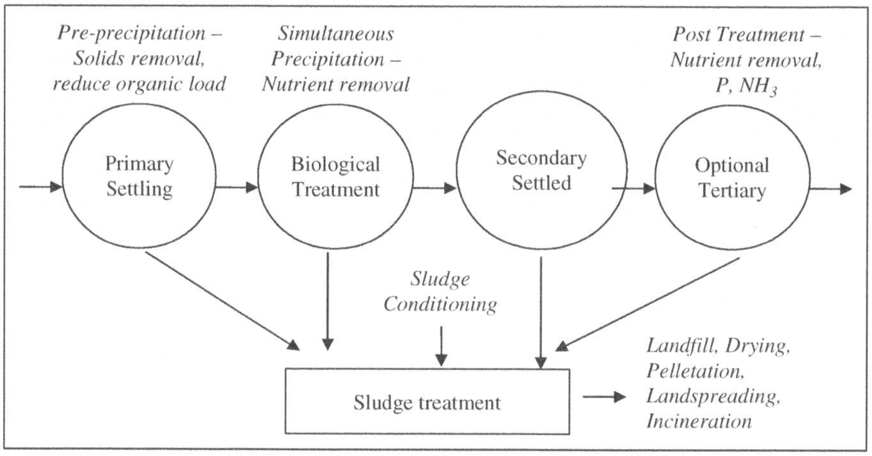

Fig. 4.14 – Examples of coagulant use in wastewater treatment

Polyelectrolyte doses for potable water clarification are generally in the range 0.1–0.5 mgl^{-1} as polymer and in the range 0.5–1.0 mgl^{-1} for wastewater. For sludge thickening, doses are normally calculated on the basis of the dry solids concentration in the sludge, and are typically in the range 1–5 kg polymer per tonne of dry solids.

Coagulation and flocculation followed by clarification can be expected to remove up to about 70% of TOC or colour and about 80% of turbidity.

4.6 Oxidation and Reduction

Section 4.6 requires a study time of about 1/2 hour

Chemical oxidation and reduction involves the transfer of electrons: an oxidising agent is an electron donor and a reducing agent is an electron acceptor (*Process Science and Engineering for Water and Wastewater Treatment* – Unit 1).

Oxidation is by far the more common of the two processes. Oxidising agents are also good disinfectants (Section 4.7) and oxidation often takes place alongside disinfection even if it is not necessarily desirable. The most common chemical oxidising agents used in water and wastewater treatment are:

- chlorine gas;
- sodium hypochlorite;
- chlorine dioxide;
- ozone;
- hydrogen peroxide;
- Fenton's reagent (hydrogen peroxide plus ferrous sulphate) and
- potassium permanganate.

The relative oxidising power of the chemicals used is indicated by their *Oxidation Potential*, the higher the potential the stronger the oxidising agent.

Table 4.1 gives some values, and indicates that hydroxyl radicals ($HO^•$) are very highly oxidising. Chemical processes dedicated to increasing hydroxyl radical levels are termed advanced oxidation processes (AOPs).

Table 4.1 – Oxidation potential values

Species	Oxidation potential (volts)
Fluorine	3.03
Hydroxyl radical	2.80
Atomic oxygen	2.42
Ozone	2.07
Hydrogen peroxide	1.78
Permanganate	1.68
Chlorine dioxide	1.57
Hypochlorous	1.49
Chlorine	1.36

Reduction reactions are comparatively uncommon and the only reducing agents of any importance are:

- sulphur dioxide;
- sodium sulphite or bisulphite;
- metallic or divalent (ferrous) iron;
- sodium borohydride or fluoroborate and
- hydrazine.

The latter two reagents are highly unstable and rarely used outside of specific industrial processes for metals recovery.

OXIDATION POTENTIAL
The higher the oxidation potential of a chemical the stronger is its oxidising capacity.

Among the common oxidation processes are:

- destruction of organic matter for colour, taste and odour control in potable water treatment;

- destruction of organic matter to reduce COD in wastewater treatment;
- destruction of ammonia in potable and wastewater;
- oxidation of pesticides in potable water treatment;
- precipitation of iron and manganese in the treatment of borehole water;
- removal of sulphides from wastewater and
- destruction of cyanide in metal finishing wastewater treatment.

Ex 4.4
What application do reducing agents serve in the disinfection of water?

Reduction reactions are uncommon but the most important are:

- dechlorination (removal of excess chlorine) in potable water disinfection;
- removal, or *scavenging*, of low levels of oxygen in boiler water treatment and
- reduction of chromates to chromium in wastewater treatment.

Sodium hypochlorite, hydrogen peroxide, Fenton's reagent, sodium sulphite and sodium bisulphite are all supplied as liquid and potassium permanganate is normally supplied as a solid for dissolution on site. These chemicals are all dosed using conventional metering pumps controlled, where appropriate, by a feedback loop with redox potential monitoring.

Chlorine and sulphur dioxide are supplied as liquefied gasses and the special gas dosing units used for these materials will be discussed in Section 4.7.

Ozone is a highly toxic gas manufactured by electric discharge of dried air, from which 2–5 volume % ozone concentration can be generated, or oxygen, which yields 7–10% ozone (Figure 4.15):

Fig. 4.15 – Ozone generator (courtesy of Anglian Water, UK)

$$3O_2 + h\nu \Rightarrow 2O_3$$

Complete systems comprise air preparation, involving driers/coolers, compressors, receivers and ozone generation by high voltage and frequency (e.g. 10 kV, 700 Hz) through a water cooled glass dielectric. All water under treatment requires contact with the ozone gas, which often means large and complex gas/liquid mixing and contact systems. The requirement of a high voltage, generated in the vicinity of water, makes safeguarding by extensive control systems necessary. The toxic nature of the gas (Threshold Limit Value –TLV 0.1 ppm) necessitates effective vent gas destructors, since it has limited solubility. Its use as a very powerful oxidising agent is tempered by its capital and operating costs.

Chlorine dioxide is a very powerful oxidising agent and, as such, is relatively unstable. Although it can be purchased as a stabilised solution, this is expensive and suitable only for small systems. More commonly, in small systems, it is prepared on site as a $15 gl^{-1}$ solution by the action of hydrochloric acid on sodium chlorite for small systems:

$$\mathbf{4HCl + 5NaClO_2 \Rightarrow 4ClO_2 + 5NaCl\ 2H_2O}$$

In larger systems a $3.5\ gl^{-1}$ solution is prepared by reacting chlorine gas with sodium chlorite.

$$\mathbf{2NaClO_2 + Cl_2 \Rightarrow 2ClO_2 + 2NaCl}$$

ADVANCED OXIDATION PROCESSES
AOPs are used to generate hydroxyl radicals ($HO^\bullet$) in order to chemically oxidise water and wastewater contaminants.

AOPs usually rely on a combination of reagents and/or equipment. The combination of ferrous ions and peroxide in Fenton's reagent generates free hydroxyl radicals:

$$\mathbf{H_2O_2 + Fe^{2+} \Rightarrow HO^\bullet + OH + Fe^{3+}}$$

Ozone and hydrogen peroxide work in combination in a similar way:

$$\mathbf{2O_3 + H_2O_2 \Rightarrow 2HO^\bullet 3O_2}$$

Combining 185nm ultraviolet (UV) irradiation with hydrogen peroxide or ozone generates free hydroxide radicals by photolysis:

$$\mathbf{O_3 + H_2O + h\nu \Rightarrow 2HO^\bullet + O_2}$$

$$\mathbf{H_2O_2 + h\nu \Rightarrow 2HO^\bullet}$$

This process is called *enhanced photo-oxidation*.

Although not strictly chemical dosing applications, there are some other AOPs which are aimed at non-biodegradable organic compounds (hard COD) in industrial wastewaters. In Wet Air Oxidation (WAO), the wastewater is heated in air to 200–300°C at 200 bar pressure (Figure 4.16 shows an example ATHOS® plant). In Supercritical Water Oxidation (SCWO), the reaction conditions are taken above the critical point of water to a temperature of 400–500°C and pressure 250–300 bar. Under these conditions most organic compounds become very soluble and are readily oxidised. Both processes are highly cost prohibitive.

Oxidation and reduction reactions are generally stoichiometric. Excess residual concentrations can be toxic to animals and potentially harmful to the environment so careful control of dosing of these materials is essential.

Fig. 4.16 – View of the ATHOS® demonstration unit in operation at the Toulouse wastewater treatment plant

4.7 Disinfection

Section 4.7 requires a study time of about 1/2 hour

Bacteria are ubiquitous. In potable water treatment there is an issue of public health in the supply of drinking water. In other industrial processes, bacteria can cause fouling of heat exchange surfaces and blockage of pipework and there is, again, an issue of public health in respect of legionellosis and similar diseases in cooling towers and air conditioning systems.

Disinfection is a chemical process, which aims to control bacterial levels by continuously dosing a chemical, often called a *biocide* or *bacteriostat*. Occasional use of high strength biocide to remove bacterial fouling (i.e. biofilms) from a closed water system is usually called *sanitisation*. Ultraviolet irradiation at 254 nm wavelength also has bacteriostatic properties and works by attacking the DNA of the bacterial cell and preventing reproduction. An example is shown in Figure 4.17.

DISINFECTION
Disinfection processes are used to control microbial growth in water systems by either continuously dosing the system with a chemical or a process.

Fig. 4.17 – UV disinfection system (courtesy of Hanovia, UK)

Biocides fall into two groups: *oxidising* and *non-oxidising*. Oxidising biocides are the chemicals, which are also used as oxidising agents, such as chlorine, chlorine dioxide, sodium hypochlorite and ozone, and these are widely used in potable water disinfection. They work by breaking the cell wall and disrupting the bacterial cell. Bromine and iodine have also been used in industrial water systems. Non-oxidising biocides are generally much more toxic – chlorinated phenols and quaternary ammonium compounds are among the chemicals which have been used – and their use is restricted to non-potable applications or closed water systems. Even here their use must be carefully controlled.

Disinfection is widely employed in water treatment:

- potable water disinfection;
- control of bacterial fouling and *legionella* in cooling towers;
- control of bacterial fouling in closed hot water and chilled water systems;
- sanitisation of high purity water distribution systems using hydrogen peroxide, peracetic acid or ozone and
- disinfection of treated sewage (Figure 4.18) or industrial effluent discharges.

Ex 4.5
What purpose does chlorine residual serve in the disinfection of water?

Fig. 4.18 – Chlorine contact tank (courtesy of Sydney Water, Australia)

Chlorine and sodium hypochlorite are, as already discussed, oxidising agents and, when used for disinfecting potable water, which contains organic matter, result in the generation of *disinfection by products* (DBPs) including *trihalomethanes* (THMs) which are tightly regulated. This has lead to a growth in the use of ozone and chlorine dioxide as primary disinfectants in potable water treatment. However, neither chemical provides a stable *residual* which is required to prevent bacterial re-growth in water distribution systems, so a small dose of chlorine is normally given as the last step in the treatment process to provide this protection.

Chlorine and sulphur dioxide are supplied as liquified gasses and must be dissolved to form a solution for dosing. Special gas dosing units, called *chlorinators* and *sulphonators*, are supplied by specialist manufacturers to control these potentially very hazardous substances.

These systems are complex and expensive because of the essential safety precautions, leak detection, drum/cylinder handling devices etc. Chlorine solution is corrosive even to stainless steel, making special materials essential; however the technology is well proven both in dosing and in closed loop *chlorine residual* control (a specific type of redox potential) and is the most widely used disinfectant system in potable water treatment. The gas is dosed into the water using an ejector which mixes intimately and produces a chlorine solution (up to about 3000 mgl^{-1}) which is carried by the pressure of the ejector to the point of application.

Safety issues with chlorine gas have made sodium hypochlorite a more popular chemical but it has short shelf life and, because hypochlorite contains only about 10% chlorine, large volumes of storage are required for large water works. Generation of hypochlorite from salt by electrolysis of sea water was developed for control of heat exchanger fouling in submarines but has now been applied as *on-site electrochlorination* for large scale production. The process obviates transportation of hazardous chlorine gas by generating the chlorine electrolytically from a brine source. Installations comprise mainly a salt saturator, base exchange softener and dosing pump for the brine preparation, a rectifier and electrolyser employing precious metal anodes for the chlorine generation stage, hypochlorite storage, hydrogen release tanks, and the usual duty and standby pumps. Because hydrogen gas is generated by the process in large volumes explosion zones must be clearly defined and correctly rated components used. Health and Safety rules dictate the installation and design requirements.

CHLORAMINATION
Chloramination is the combining of chlorine and ammonia to produce chloramines for disinfection.

In some circumstances chlorine can be combined with ammonia to produce chlorinated ammonia compounds called *chloramines*. Chloramines, NH_2Cl (monochloramine) and $NHCl_2$ (dichloramine), generally have a lower bactericidal potency than chlorine but are chemically more stable. This process, called *chloramination*, is becoming more widely used in potable water treatment due to its stability in the distribution network.

Ozone is a powerful oxidant that is capable of oxidising organic matter carbon dioxide and water without generating DBPs. However, its lack of by-product generation is tempered by its high capital and operating costs, the associated hazards and the possibility of bromate formation. It is, however, gaining ground in pharmaceutical systems where its instability is an advantage in that it leaves no residual. In these rather small scale applications it is generated in solution by electrolysis of water in a special cell.

Ex 4.5
What are the differences in the application off chloramines compared to chlorine in water disinfection?

Chlorine dioxide is used in many food industry and other industrial applications. Although it does not form THMs, its use is restricted in potable water applications because chlorite formation occurs at doses above about 0.5 mgl^{-1}.

Typical disinfectant doses vary from around 0.1–0.5 mgl^{-1} of chlorine or ozone for disinfection of clean potable water up to 5 mg/l or even more for highly contaminated process and wastewaters.

Sanitisation of pharmaceutical systems typically uses 0.2–0.5 mgl^{-1} of ozone and sanitisation of semiconductor ultrapure water systems may use 2% hydrogen peroxide or peracetic acid.

Because of the possibility of side reactions with oxidisible substances in the water it is usually advisable to carry out a *chlorine demand* test on a water sample before specifying a chlorine dose.

We can compare disinfectant power by looking at *specific lethalities* which we shall discuss in more detail in a future module. Specific lethalities for the main disinfectants for *E.coli* and for poliovirus are given in Table 4.2.

Table 4.2 – Specific lethality (L) values*

Disinfectant	*E. coli*	Poliovirus 1
Ozone	2300	920
Chlorine	120	4.6
Chlorine dioxide	16	2.4
Hypochlorite	5	0.5

*$\mathbf{ln\ (N/N_0) = -\ L.t.C^n}$

where: N, N_0 = number of organisms originally & at time t
C = disinfectant concentration
n = coefficient of dilution
L = takes units of $(l/mg)^{1/n}$ per unit time

Section 4.8 requires a study time of about 1/4 hour

4.8 Scale and Corrosion Inhibition

Many water systems are prone to scale and corrosion (Figure 4.19). Corrosion can, of course, damage pipework and steel process plant like heat exchangers and the corrosion products can cause fouling of heat transfer surfaces, pipes and valves. Corrosion also results in soluble metal ions contaminating the water which, in the case of copper, zinc or lead can cause problems in drinking water. In industrial systems chemical corrosion inhibitors like zinc/sulphate or nitrite/bromate blends can be applied. In potable water, where these are not permitted, it is more usual to adjust the Langelier Saturation Index (*Process Science and Engineering for Water and Wastewater Treatment* – Unit 2) by adding alkali or calcium. In the case of lead, the addition of phosphoric acid to provide 1 mgl^{-1} of phosphate in the water has been found to inhibit *plumbosolvency*.

PLUMBOSOLVENCY
Plumbosolvency is the suppression of lead concentrations in water supplies using orthophosphate dosing.

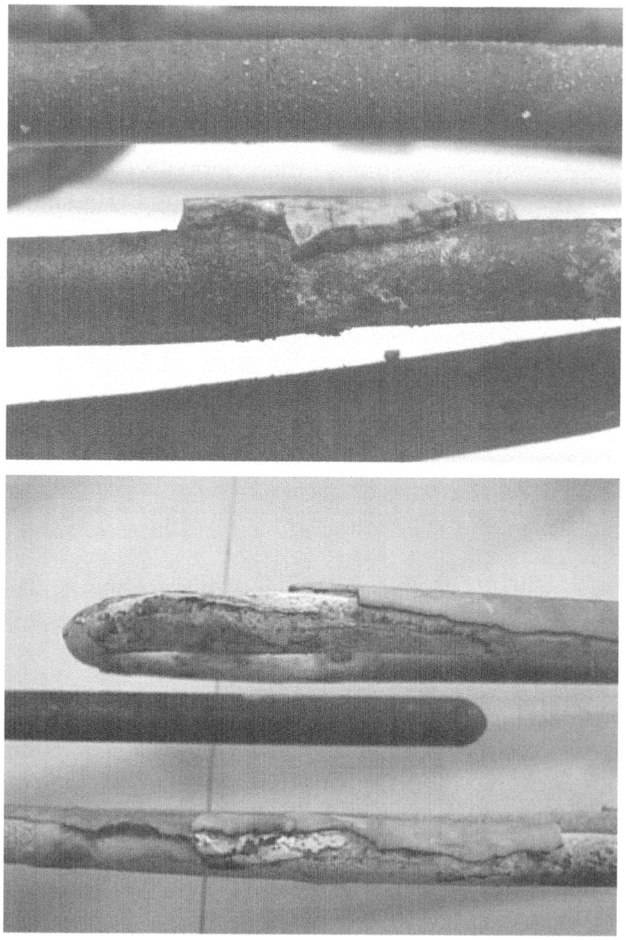

Fig. 4.19 – Examples of scaling and corrosion on heater elements

Scaling and corrosion are problems faced in many industrial applications but also in reverse osmosis and distillation systems. Sodium hexametaphosphate (*Calgon*) was one of the earliest scale inhibitors but this has now been superseded by a range of synthetic organic chemicals including phosphonates, acrylates and acetates. Not all of these have approval for use in potable water treatment.

Corrosion inhibitors are commonly used in industrial cooling water systems. Hot water systems are generally less prone to corrosion because heating reduces the solubility of oxygen, the presence of which is essential for corrosion to occur.

Typical examples include:

- open evaporative cooling towers;
- chilled water circuits;
- water distribution systems for prevention of *red water* and
- water distribution systems for plumbosolvecny suppression.

Scale inhibitors are added in many applications including:

- open evaporative cooling towers;
- chilled water circuits;
- hot water boilers;
- low pressure steam boilers;
- reverse osmosis feed and
- distillation systems.

It is difficult to measure corrosion rates on-line and many of the chemicals dosed are not amenable to on-line analysis. Consequently most corrosion and scale inhibiting chemicals are simply dosed in proportion to flow – typically in proportion to *make-up* flow in cooling towers and boilers. Where on-line analysis is available (e.g. for phosphate in plumbosolvency suppression) the measurement is usually used for monitoring only and not for feedback control.

4.9 Self Assessment Questions

SAQ 4.1 Discuss the main factors that effect the formation of flocs in water treatment.

SAQ 4.2 Explain how the coagulation/flocculation process is influenced by typical additives used in practice.

SAQ 4.3 For the removal of what contaminants in the treatment of potable water would you use coagulation/flocculation?

Why are aluminium and iron salts particularly effective coagulants for potable water treatment?

SAQ 4.4 Identify which of the following chemicals/chemical processes (i–iv) that:

a) react(s) with ammonia
b) generate(s) no, or only low levels, of THMs
c) is/are capable on inactivating viruses
d) has/have a disinfection capability which varies significantly with pH
e) generates hydoxyl radicals
f) is/are termed (an) "AOP(s)"
g) can be produced by the combination of chemical
h) is/are (a) reducing agent(s)
i) normally incurs the highest capital cost
j) can be generated both in situ and provided as a stabilised solution.

i) Chlorine
ii) Fenton's reagent
iii) Ozone
iv) Chlorine dioxide

SAQ 4.5 Which of the above chemicals/chemical processes (i–iv) is normally used for pesticide removal in potable water treatment?

What downstream process is it invariably combined with and why?

SAQ 4.6 Which of the above chemicals/chemical processes (i–iv) is almost always used as the final unit operation in potable water treatment and for what principal reason?

SAQ 4.7 The same chemicals/chemical processes (i–iv) is also used for "breakpointing".

What is this process and what products and intermediates are formed by the chemical reaction involved?

4.10 Solutions to Exercises

Ex 4.1 Chemical mixing can be achieved by static or rapid flash mixing.

Static or hydraulic mixers induce turbulence by placing blades into the pipe or jumps into an open channel, which results in turbulence.

Rapid flash mixing using some sort of mechanical mixer, typically a propeller, turbine or paddle, which provides a high shear gradient for a residence time of about 1–2 minutes.

Ex 4.2 The pH of the solution is important because not all metal show the same relationship between pH and solubility. e.g. aluminium hydroxide is soluble at low or high pH whereas iron precipitates with increasing pH.

Ex 4.3 Jar testing is typically carried out to optimise and establish coagulant and pH conditions for removing colloids in water and wastewater treatment.

The optimum coagulant concentration and pH are determined in order to neutralise the surface charge on the particle so that the particles can bind to a settleable size.

Ex 4.4 Reducing agents such as sulphur dioxide; sodium sulphite or bisulphite are used for dechlorination (removal of excess chlorine) in potable water disinfection.

Ex 4.5 Chlorine residuals provide a stable residual, which is required to prevent bacterial re-growth in water distribution systems.

Ex 4.6 Chloramination is the combination of chlorine and ammonia to produce chloramines. Chloramines generally have a lower bactericidal potency than chlorine but are chemically more stable and are becoming more common in water disinfection.

Bibliography

Unit 4 – Chemical Processes

Droste R.L. *Theory and Practice of Water and Wastewater Treatment*. John Wiley and Sons Inc., Chichester (1997).

Kawamura S. *Integrated Design of Water Treatment Facilities*. 2nd ed., John Wiley and Sons, Inc., New York (2000)

Metcalf and Eddy, *Wastewater Engineering, Treatment and Reuse*. 4th ed., McGraw Hill, New York (2003).

Parsons S.A. *Advanced Oxidation Processes for Water and Wastewater Treatment*. IWA Publishing, London (2003)

Sincero A.P. and Sincero G.A. *Physical-chemical Treatment of Water and Waste water*. IWA Publishing, London (2002).

5. Sorption Processes

Contents

Aims and Objectives

This Unit covers the principles of sorption processes used in water and wastewater treatment:

After studying these notes you should be able to:

1. explain in your own words the following terms and concepts:

   adsorption
   ion exchange
   fixed bed
   softening
   deionisation

2. describe the cyclic operation of sorption processes and the need for sorbent regeneration.

It is important that you are able to complete all the self assessment questions at the end of this Unit.

Essential Prerequisites

It is not necessary to have a completed any other Units before undertaking this Unit.

Additional, information on the fundamentals for this Unit, refer to the following Units in *Process Science and Engineering for Water and Wastewater Treatment*:

Unit 1 Fundamentals of Water Chemistry
Unit 2 Chemical Kinetics and Equilibria
Unit 7 Fundamentals of Process Engineering
Unit 11 Engineering Hydraulics

5.1 Introduction

In both water and wastewater treatment a wide range of organic materials are amenable to removal by sorption. In water treatment, taste and odour compounds or "off-flavours", pesticides and other disinfection by-products from chlorination can be removed by simple attachment to the surface of a solid substrate. In wastewater treatment, the intermediate addition of sorbent materials is used to improve sludge settleability and remove toxic compounds.

In this Unit the principles involved in the sorption of organic compounds onto the surface of solid substrates and their application in different water and wastewater unit processes are outlined.

Section 5.2 requires a study time of about 1/2 hour

5.2 Sorption

Sorption is a separation process in which dissolved material (*sorbate*) is removed from the liquid phase onto another, usually solid phase (*sorbent*). The sorbate is held onto the sorbent by physical forces (*adsorption*) or chemical reaction (*chemisorption*).

Classical sorption theory provides a model in which the surface of the sorbent has a fixed number *active sites* each of which is capable of sorbing a single molecule of sorbate. It follows that a given mass of sorbent will have a fixed capacity (dependent on its surface area) which can take up a given number of molecules. The number of molecules multiplied by the molecular weight is the *capacity* of the sorbent in g of sorbate/g sorbent.

SORPTION
Sorption is a physical process where soluble molecules are removed by their attachment to a solid substrate.

When each site is occupied the sorbent is said to be *exhausted* and, depending on the nature of the process, it may be possible to restore the capacity or *regenerate* the sorbent. So sorption can be considered, like filtration, as a *batch process* operating in a two part cycle: exhaustion and regeneration.

In water treatment, adsorption is widely used for the removal of organic matter, the adsorbent being most commonly employed is activated carbon, although some other adsorbents, like activated alumina, clay colloids, hydroxides and adsorbent resins are also available. In adsorption processes the molecules of adsorbate are retained on the surface of the adsorbent by surface forces:

- van der Waals forces and
- hydrogen bonding.

Ex 5.1
Describe what is meant by a sorption process becoming exhausted?

Chemisorption is commonly used for the removal of ionic species in industrial water purification and, occasionally, for industrial effluent treatment. The chemisorbents are almost invariably synthetic *ion exchange resins* although a few natural minerals, called *zeolites*, are used for special purposes.

Most sorbents are *porous* media (*Process Sciences and Engineering for Water and Wastewater Treatment* – Unit 11) and impose a hydraulic restriction (*head loss)* on the water flowing through them. Typical head losses in sorption processes are in the range 1–10 m (0.1–1bar).

5.3 Adsorption

Section 5.3 requires a study time of about 1 hour

Adsorption is a surface phenomenon so one of the most important characteristics of an adsorbent are its affinity for organic molecules and its surface area. In some industrial applications, special high porosity ion exchange resins called *organic scavengers* are used for the adsorption of organic matter. However, in almost all other applications, activated carbon is the most widely used adsorbent.

Activated carbon is a highly porous material with surface area of around 1000 m^2/g. It is manufactured from a variety of carbonaceous raw materials, such as wood, lignin, coal, bone and coconut shells. The first step in manufacturing is *carbonisation* that is pyrolysis at fairly high temperatures in the absence of air to burn-off of non-carbon impurities, leaving a porous structure. The resulting *charcoal* is then *activated* by controlled oxidation at a temperature of 800° C or higher, sometimes with chemical activators, to give a structure with very fine pores (*micropores* – up to about 10 nm) and some larger pores (*macropores* – up to about 10 mm).

ACTIVATED CARBON
Activated carbon is the most widely used adsorbent which is used either as powder (PAC) or as granules (GAC).

Most of the surface area and adsorption capacity is associated with micropores, but macropores are very important in providing access to adsorption sites (i.e. give a more rapid rate of adsorption).

Adsorption is used principally to remove organic matter from water and, occasionally, from wastewater. As such duties include:

- removal of taste and odour from drinking water;
- removal of colour from drinking water;
- removal of THM precursors from drinking water;
- removal of pesticides (either alone or after ozonation);
- removal of natural organic matter to protect downstream processes in industrial water purification;
- dechlorination;
- de-ozonation;
- removal of solvents from industrial wastewater and
- removal of COD from industrial wastewater.

Activated carbon is used in two modes, either as *granular activated carbon* (GAC) or as *powdered activated carbon* (PAC). Figure 5.1 show the typical application options for GAC and PAC at a potable water plant.

GRANULAR ACTIVATED CARBON
GAC is used in continuous and batch applications in either packed or expanded beds.

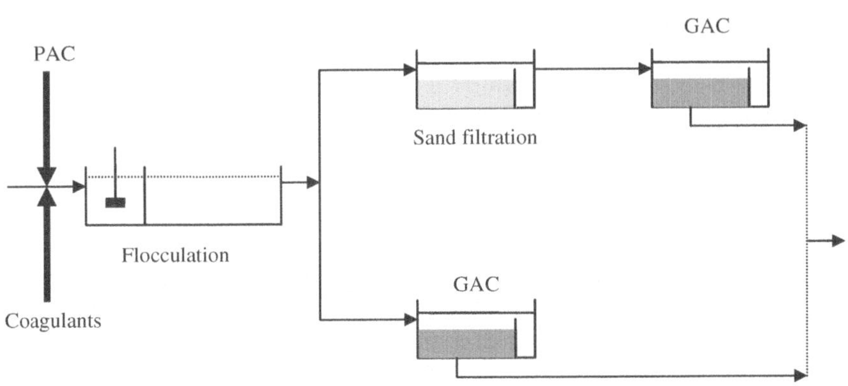

Fig. 5.1 – Application of activated carbon in water treatment

GAC is manufactured in the form of granules about 0.5–1.5 mm in diameter (Figure 5.2) and is used in expanded *fixed beds* like gravity sand filters through which the water passes down through the activated carbon. The GAC process can be employed a single sorption system or as a secondary process following sand filtration (Figure 5.1). Organic molecules such as pesticides are adsorbed from the water as it passes through the bed and the carbon becomes exhausted from the top downwards. More recently, GAC processes have been allow to develop or *ripen* into biofiltration systems, whereby microbial biofilms develop on the surface of the GAC aiding in the biodegradation of the sorbed organics.

Ex 5.2
Which activated carbon would be most appropriate for the periodic removal of algal by-products and why?

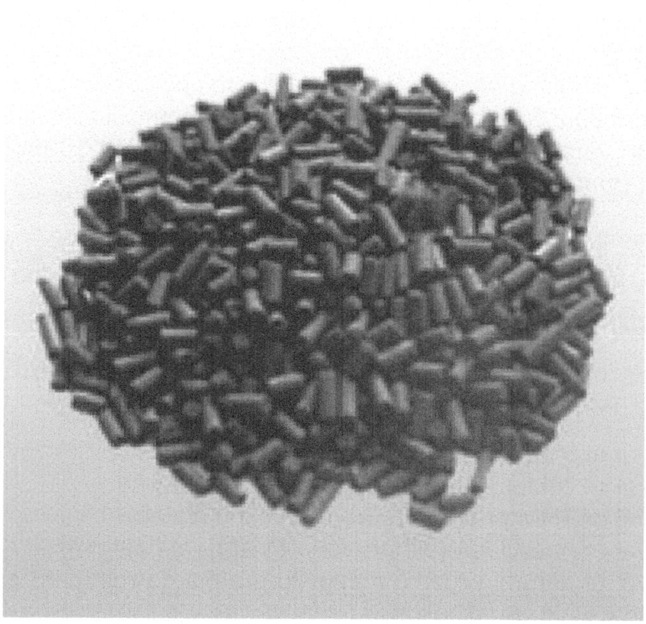

Fig. 5.2 – Granular activated carbon

When the carbon becomes exhausted it is either disposed of by incineration or regenerated off-site, usually by re-activation in a rotary hearth furnace. In wastewater treatment, solvents may be recovered by regenerating the carbon with high pressure steam to volatilise the solvents, which can then be recovered by condensation.

Industrial GAC filters use the same sort of packed bed arrangement employed in sand pressure filters (Unit 3). Figure 5.3 shows a typical example of such as GAC packed filter at a water treatment plant. The plant treats 30 Mld of water for pesticide removal and consists of eight pressure vessels, each 3.3 m diameter and each charged with 20 m^3 of granular activated carbon.

PAC is produced as a fine powder and is dosed into the water to be treated as a slurry, intimately mixed and then removed by coagulation, sedimentation and filtration and disposed of with the sludge (Figure 5.1). Although it is very expensive it is frequently used for treating drinking water supplies from sources, which have high levels of natural organic matter (NOM) or seasonal occasions of algal metabolites such as geosmin. This type of intermediate application is only for short periods of the year and is generally more cost effective than GAC because it does not require the very high capital cost of a permanent fixed bed installation.

Fig. 5.3 – GAC treatment of water

The principal parameter in adsorption operations is the *empty bed contact time* (EBCT), defined as the volume of adsorbent divided by the flow rate. The EBCT for most water treatment applications is around 10–20 minutes. As long as the carbon at the bottom of the bed is not exhausted, removal of the organic material is typically better than 99%.

Adsorption capacities are high: typical commercial activated carbons can adsorb up to about 20% of their weight of organic substances from water.

EMPTY BED CONTACT TIME
EBCT is defined as the volume of absorbent by the flow rate.
EBCT $=$ V/Q h
Where:
V $=$ volume of GAC (m^3)
Q $=$ flowrate ($m^3\ h^{-1}$)

Section 5.4 requires a study time of about 1 hour

5.4 Ion Exchange Softening

Ion exchange softening is the simplest and commonest of the chemisorption processes. The sorbent (in this case a *strongly acidic cation exchange resin*) is in the form of 0.5–1.0 mm diameter beads of polystyrene. As in adsorption processes, water to be treated passes downwards through a deep bed of the resin. The active sites on the ion exchange resin beads are called *functional groups* and, in this case, they are anionic *sulphonic* groups $-SO_3^-$. The sulphonic groups can take up cations like sodium, calcium and magnesium, but have a preference for di- and tri-valent ions rather than monovalent. If water containing hardness salts of calcium and magnesium is passed through a bed of ion exchange resin *in the sodium form* then the calcium and magnesium ions in solution will be chemisorbed onto the resin and the sodium ions from the resin will go into solution. Thus the calcium and magnesium ions are exchanged for sodium. For example, softened water for industrial use prevents the formation of scum or scale deposits. Typical applications for ion exchange water softeners include:

Ex 5.3
Which metal ions would you expect to be removed by a sodium cycle softner?
a) Ferrous iron
b) Potassium
c) Barium
d) Sodium

- laundries;
- cooling tower make-up;
- low pressure steam boilers;
- hot water calorifiers (heaters);
- domestic hot and cold water services;
- glass washing and
- bottle washing in dairies.

Because calcium and magnesium ions are removed in exchange for sodium, the sodium concentration of the water is increased by softening and may exceed the drinking water quality specification for sodium, so the process is not recommended for drinking water supplies.

During the *exhaustion* phase (or *service run*), that typically lasts from 8–24 hours, calcium and magnesium ions are removed from solution in exchange for sodium. The reaction is as follows:

$$\mathbf{2R - SO_3Na} + \mathrm{Ca^{++}} \Leftrightarrow \mathbf{(R - SO_3)_2Ca + 2Na^+}$$

In this reaction R represents the polystyrene *resin matrix*. Note that the reaction is an equilibrium reaction so (*Process Sciences and Engineering for Water and Wastewater Treatment* – Unit 1) it is reversible. During *regeneration* a strong *regenerant* solution of sodium chloride is passed through the resin bed and discharged to drain. The high concentration of sodium forces the equilibrium in the reverse direction:

$$\mathbf{(R - SO_3)_2Ca + 2NaCl \Leftrightarrow 2R - SO_3Na + CaCl_2}$$

The waste stream or *spent regenerant* contains all the calcium and magnesium which was removed during the service run plus excess regenerant, which is typically 40–60% of the applied regenerant.

Small commercial softeners (Figure 5.4) are generally constructed using GRP vessels to contain the resin with *multiport* control valves which carry out the regeneration sequence of *injecting* the regenerant solution into the bed and then *displacing* it and *rinsing* the bed before returning the softener to service.

ION EXCHANGE SOFTENING
Active sites on an ion exchange resins adsorb di- and tri-valent ions in exchange for mono-valent ions like sodium.

Ion exchange softening is typically applied to waters of total hardness up to about 10 meq/l (*Process Sciences and Engineering for Water and Wastewater Treatment* – Unit 2) and removes virtually all of this leaving a *residual hardness* of about 0.1 meq/l.

Fig. 5.4 – Commercial water softener

Ex 5.4
Which of the following salts could be used for regenerating a sodium cycle water softener?
a) Potassium chloride
b) Magnesium chloride
c) Hydrochloric acid
d) Sodium nitrate

Resin *capacities* are normally expressed as either equivalents or mass (usually grammes) as calcium carbonate of hardness removed per unit volume (usually litre) of resin. The amount of regenerant chemical applied, the *regeneration level*, is usually expressed as mass of 100% chemical (usually grammes) or less commonly as equivalents of regenerant chemical per unit volume. Expressing both the regeneration level and capacity in equivalent units allows the calculation of a dimensionless number, the *regeneration ratio* which is an indication of how much of the applied chemical is unused. A regeneration ratio of 1 indicates stoichiometric efficiency: a regeneration ratio of 2 indicates that twice the stoichiometric quantity of regenerant chemical is being used or that 50% of it is wasted.

Softening resins have a capacity for hardness removal of about 1 eql^{-1} and are normally regenerated at a regeneration level of about 120gNaCl/l (about 2 eql^{-1}) giving a regeneration ratio of 2.

Section 5.5 requires a study time of about 1/2 hour

5.5 Ion Exchange Deionisation

If the strongly acidic cation exchange resin used in the softening process is regenerated not with salt but with hydrochloric acid, the sulphonic groups will take up hydrogen ions and the resin will operate *in the hydrogen cycle*. The resin will chemisorb all cations in preference to hydrogen, which is released in exchange. The resulting *decationised* water contains only hydrogen ions and anions and is, therefore a mixture of mineral acids. If this decationised water is passed through a bed of *strongly basic anion exchange* the anions can be chemisorbed in exchange for hydroxyl. The strongly basic resin has cationic quaternary ammonium functional groups $-(NCH_3)_3^+$ which can be regenerated with caustic soda to operate in the hydroxyl cycle.

The resulting deionised water is widely used in industry for, among other applications:

- high pressure boiler feed;
- general laboratory uses;
- double glazed glass washing;
- fine chemicals manufacturing;
- humidifiers;
- metal finishing and
- battery top up.

Water to be deionised is passed through a bed of strongly acidic cation exchange resin in the hydrogen form which exchanges all cations for hydrogen:

$$\mathbf{R-SO_3H + NaCl \Leftrightarrow R-SO_3Na + HCl}$$

Regeneration is with hydrochloric acid:

$$\mathbf{R-SO_3Na + HCl \Leftrightarrow R-SO_3H + NaCl}$$

The water then passes through a bed of strongly basic anion exchange resin in the hydroxyl form which exchanges all anions for hydroxyl:

$$\mathbf{R-N(CH_3)_3OH + HCl \Leftrightarrow R-N(CH_3)_3Cl + H_2O}$$

Regeneration is with caustic soda:

$$\mathbf{R-N(CH_3)_3Cl + NaOH \Leftrightarrow R-N(CH_3)_3OH + NaCl}$$

Ex 5.5
What would you expect to be the pH of water leaving a hydrogen cycle cation exchanger prior to anion exchange?
a) 7.0
b) 2.0
c) 8.5

Industrial two-bed deionisers (Figure 5.5) are designed so that all waste regenerant streams contain all the salts removed from the water during the service run plus excess regenerant acid and caustic soda.

Cation resins have a capacity of removal of about 1 eql^{-1} and are normally regenerated at a regeneration level of about 65 gHCl/l (about 1.8 eql^{-1}) giving a regeneration ratio of 1.8. Anion exchange resins are less efficient and have a capacity of about 0.6 eql^{-1} at a regeneration ratio of about 2.5. Their kinetics is also slower than those of cation resins and a maximum hydraulic loading of about 30 BVh^{-1} applies for anion resins.

Fig. 5.5 – Industrial two-bed deioniser

Section 5.6 requires a study time of about 1/2 hour

5.6 Mixed Bed Polishing

The quality of two-bed deionisation is limited by equilibrium conditions, mixed bed polishing overcomes this limitation by continuous removal of the reaction products.

Mixed bed polishers are used downstream of two bed deionisers where higher quality water ($<0.1\mu$ Scm^{-1} conductivity) is required. Applications include:

- power station boilers;
- pharmaceuticals and cosmetics manufacturing;
- analytical laboratory use and
- semiconductor manufacturing.

MIXED BED POLISHING
Mixed bed polishing is used downstream of two bed deioniser where high quality water is required.

Deionised water is passed through a bed of ion exchange resin containing, typically, half strongly acidic cation exchange resin and half strongly basic anion exchange resin, respectively in the hydrogen and hydroxyl forms. Both deionisation reactions occur (Section 5.5) because the resins are in an intimate mixture. The products of *salt splitting* reactions are immediately removed by adjacent resin beads to form water:

$$\mathbf{R - SO_3H + NaCl \Leftrightarrow R - SO_3Na + HCl}$$

$$\mathbf{R - N(CH_3)_3OH + NaCl \Leftrightarrow R - N(CH_3)_3Cl + NaOH}$$

$$\mathbf{R - SO_3H + NaOH \Leftrightarrow R - SO_3Na + H_2O}$$

$$\mathbf{R - N(CH_3)_3OH + HCl \Leftrightarrow R - N(CH_3)_3Cl + H_2O}$$

This means that the reactions are effectively no longer reversible since water remains essentially associated with only very low levels of the dissolved acid (H^+) and hydroxide (OH^-) ions. The water is thus not in equilibrium with the resin and leakage is almost completely eliminated.

In small systems and in the production of *ultrapure water* the polishing mixed bed resins are retained in simple polypropylene lined GRP vessels. When the resins are exhausted (typically after 6–12 months) they are taken off site for regeneration by a third party supplier or are simply thrown away and replaced with new resins. In large power station plant this would not be economic and the resins are regenerated in situ by separating them by backwashing, regenerating the cation resin with acid and the anion resin with caustic soda and then re-mixing them by blowing air through the bed.

A well designed and operated polishing mixed bed can easily meet a conductivity of 0.1 μScm^{-1} and the process is vital to the production of ultrapure water with conductivity <0.56 μScm^{-1}.

5.7 Ion Exchange in Wastewater Treatment

Section 5.7 requires a study time of about 1/2 hour

Ion exchange can be used for the treatment of industrial wastewaters, which contain ionic contaminants and a wide range of special resins is available for specific purposes.

Among some of the applications for ion exchange in wastewater treatment are:

- recycling of metal plating rinse water;
- removal of heavy metals from plating waste streams;
- removal of mercury from chemical process wastewaters and
- recovery of precious metals.

Conventional two stage deionisation is used for the preparation and recycling of rinse waters in the plating industry but the use of weakly basic anion exchange resins with tertiary amine functionality $-N(CH_3)_2$ offer cost advantages because they have higher capacity and lower regeneration ratio.

Weakly acidic cation exchange resins with carboxylic functionality $-COO^+$ have a higher affinity for divalent metals than for monovalent and can be used to remove metal ions from wastewaters containing sodium. A strongly acidic sulphonic resin would remove sodium ions as well even if this is unnecessary.

Cation exchange resins with imminodiacetate functionality $-CH_2.N(CH_2.COO)_2{}^+$ have a very high affinity for heavy metals like mercury, copper, lead, nickel, zinc and cadmium over monovalent and even alkaline earth metals – calcium, magnesium, barium – so can be used for removing trace levels of these metals from treated wastewaters.

These special resins are very expensive and regeneration procedures are often very complicated. However, their superior chemical performance in terms of targeting specific contaminants often makes them cost effective. Correctly applied, they are capable of reducing low heavy metal concentrations to less than 10 μgl^{-1}. Performance in particular applications is not easy to predict and laboratory tests are advisable.

5.8 Self Assessment Questions

SAQ 5.1 What sorption process would you select for the following duties:

a) removal of pesticides from drinking water
b) removal of hardness from cooling water
c) removal of chlorine from mains water

SAQ 5.2 What are appropriate values for the regeneration ratios for SAC and SBA resins in two-bed deionisation?

SAQ 5.3 Give four examples of contaminants for the removal of which you would consider activated carbon adsorption.

How is granular activated carbon regenerated?

What are the relative merits and applications for granular and powdered activated carbon?

SAQ 5.4 Describe the two bed ion exchange deionisation process.

Why does a mixed bed deioniser produce treated water of higher quality that a two bed deioniser?

5.9 Solutions to Exercises

Ex 5.1 A sorption process becomes exhausted when its active sorption sites are all occupied, it may be possible to restore the capacity or regenerate the sorbent.

Ex 5.2 Powdered activated carbon is the most suitable process where treatment is not required all year round because PAC dosing can be easily stopped when not required thus reducing operating costs.

Ex 5.3 Ferrous iron and Barium are both divalent ions and are removed efficiently by sodium cycle softening. Potassium is a monovalent ion and will not be removed.

Ex 5.4 Potassium chloride would also work as a softener regenerant although the operating cycle would then be potassium rather than sodium and, again, very expensive.

Sodium nitrate will supply the necessary sodium ions (although more expensive than sodium chloride) so could be used for regeneration.

Ex 5.5 *2.0.* The cation exchanger removes all cations from the water in exchange for hydrogen ions so that the water leaving the cation unit will contain a mixture of mineral acids depressing the pH.

Bibliography

Unit 5 – Sorption Processes

Gray N. F. *Water Technology*. Arnold, London (1999).

Montgomery J. *Water Treatment Principles and Design*. John Wiley Interscience, New York (1997).

Morelli C.D. *Basic Principles of Water Treatment*. Tall Oaks Publishing, Littleton (1996).

Owens D.L. *Practical Principles of Ion Exchange*. 2nd ed., Tall Oaks Publishing, Littleton (1995).

6. Biological Processes

Contents

Aims and Objectives

This Unit covers the principles of biological processes used in water and wastewater treatment works:

After studying these notes you should be able to:

1. explain in your own words the following terms and concepts:

   aerobic processes
   anaerobic processes
   fixed film
   suspended growth

2. know the main types of biological processes

   trickling filter
   rotating biological contactors
   biological aerated filters
   fluidised beds
   activated sludge
   sequencing batch reactors
   membrane bioreactor
   anaerobic digestors
   anaerobic filters,
   anaerobic fluidised beds
   upflow anaerobic sludge blanket reactors

3. describe the different parameters used to monitor a biological treatment process.

It is important that you are able to complete all the self assessment questions at the end of this Unit.

Essential Prerequisites

It is not necessary to have a completed any other Units before undertaking this Unit.

Additional, information on the fundamentals for this Unit, refer to the following Units in *Process Science and Engineering for Water and Wastewater Treatment*:

Unit 4 Fundamentals of Microbiology
Unit 5 Fundamentals of Biochemistry
Unit 10 Reactor Design Theory

6.1 Introduction

Biological treatment of wastewaters rely upon micro-organisms that utilise the dissolved and colloid organic matter as food. They can often offer an economic alternative to other secondary chemical treatment processes. Biological processes are usually applied after primary treatment, as the wastewater still contains significant amounts of organic matter that needs to be removed before discharge.

In this Unit the principle processes involved in the biological conversion of dissolved and colloid organic matter using microbial biomass that can be readily separated from the effluent are outlined.

Section 6.2 requires a study time of about 1 hour

6.2 Biological Processes

Biological treatment processes are used to some extent in the production of potable water – for iron and manganese removal, nitrate removal and pesticide removal – but it is in the treatment of municipal and high strength industrial wastewater treatment that they have become ubiquitous. They are based on the simple concept of micro-organisms consuming soluble and colloidal organic compounds (measured as *Biochemical Oxygen Demand, BOD*) present in the wastewater as a nutrient source. The micro-organisms grow and are subsequently separated from the wastewater which has had the BOD removed. These micro-organisms derive energy and cellular material from the oxidation of this organic matter and can be aerobic (need oxygen) or anaerobic (need absence of oxygen) (*Process Science and Engineering for Water and Wastewater Treatment* – Unit 4). Biological wastewater treatment is also capable of removing other wastewater components besides organic matter, including suspended solids, ammonia, heavy metals and xenobiotics.

Advantages of biological processes are that:

- they are "natural";
- wastes are converted to gases;
- the sludge produced is treatable;
- one of the waste products (methane gas) can be used as an energy source;
- odours are usually reduced;
- a high removal efficiency possible;
- variable loads can be tolerated and
- constituents other than BOD are removed.

Disadvantages are that these processes are:

- susceptible to toxic chemicals;
- slow compared to chemical processing;
- generate solids that need to be disposed of;
- produce noxious compounds;
- produce aerosols and
- energy consuming (if aerobic).

6.2.1 Microbiology

Many species and genera from the kingdom of protists (*Process Science and Engineering for Water and Wastewater Treatment* – Unit 4) have a significant role in biological treatment systems including:

Ex 6.1
What is the source of micro-organisms in biological wastewater treatment systems?

Bacteria

- conversion of soluble and particulate organic compounds into biomass and gaseous waste products (CO_2 and CH_4).
- conversion of ammonia to nitrate (nitrification).
- conversion of nitrate to nitrogen gas (denitrification).
- conversion of soluble phosphate into insoluble intracellular phosphate (permitting removal from dischargeable effluent).

Protozoav

- consume particulate organics (including bacteria).
- play an important role in the removal of suspended solids.

Fungi

- may assist bacteria in the removal of organics in trickling systems.

Algae

- role in nutrient uptake in specialised tertiary treatment systems.

Others

- larger biological species such as rotifers, nematode worms and insect larvae may contribute to the consumption of particulate organic matter, especially in trickling filter systems.

In most biological systems the bacteria play the main role in the primary reduction of contaminants, but grazing organisms are also important in maintaining a balanced microbial ecosystem. The genera responsible for significant processes within wastewater treatment are shown in Table 6.1 and can be classified by the required redox conditions into three types: *obligate aerobes* (1); *facultative anaerobes* (2) and *obligate anaerobes* (3) as shown in Table 6.2 (*Process Science and Engineering for Water and Wastewater Treatment* – Unit 4).

BIOLOGICAL SYSTEMS
Bacteria are the predominate micro-organisms in biological wastewater treatment.

Table 6.1 – Bacteria in wastewater treatment

Genus	Function
Pseudomonas	Removal of carbohydrate, slime production and denitrification
Zoogloea	Slime production, floc formation
Bacillus	Protein and starch degradation
Arthrobacter	Carbohydrate degradation (1)
Microthrix	Fat degradation, filamentous growth
Acinetobacter	Phosphorus uptake and release
Nitrosomonas	Nitrification of ammonia to nitrite (1)
Nitrobacter	Nitrification of nitrite to nitrate (1)
Achromobacter	Denitrification (2)
Clostridium	Anaerobic hydrolysis of biochemicals (3)
Desulfovibrio	Sulphate reduction (3)
Methanothrix	Methane production (3)

Table 6.2 – Types of microbial metabolism used in biological wastewater treatment processes

Component	Process	Electron acceptor	Type
Organic – carbon	Aerobic biodegradation	O_2	1
Ammonia	Nitrification	O_2	1
Nitrate	Denitrification	NO_3^-	2
Sulphate	Sulphate reduction	SO_4^{2-}	3
Organic – carbon	Methanogenesis	CO_2	3

Those bacteria responsible for organic (both dissolved and colloidal matter) removal and denitrification are *heterotrophs*, i.e. organic carbon is required as an energy source and as the carbon source. Those responsible for nitrification, sulphate reduction and methanogenesis are *autotrophs*, i.e. inorganic chemical reactions provide the energy and inorganic carbon is needed as the carbon source (*Process Science and Engineering for Water and Wastewater Treatment* – Unit 4).

6.2.2 Aerobic and Anaerobic Processes

BIOLOGICAL PROCESSES
Biological treatment processes are either aerobic (need oxygen) or anaerobic (need absence of oxygen).

Although bacteria are found across a wide spectrum of redox conditions in wastewater treatment, the two major groups of treatment systems are aerobic and anaerobic processes. In general, anaerobic catabolism produces less energy than aerobic catabolism; therefore larger reactor volumes are required. The advantages and disadvantages of the two processes are summarised in Table 6.3.

Table 6.3 – Aerobic versus anaerobic systems

	Aerobic	Anaerobic
Energy	High	Low
% removal	95+	60–90
Sludge production	High	Low
Stability	High–moderate	Low–moderate
Start-up	2–4 weeks	2–4 months
Odour	Low	Potential
Alkalinity	Low	High
Biogas production	No	Yes
Nutrients	Can be high	Low

BIOLOGICAL PROCESSES
Biological treatment processes are either aerobic (need oxygen) or anaerobic (need absence of oxygen).

Anaerobic processes are suitable for high strength wastewaters, i.e. high BOD. This also includes sludge high in organic solids. Anaerobic treatment is often used for the digestion of sewage sludges resulting in the reduction of solids, pathogen destruction, odour removal and finally methane production. Some anaerobic metabolic pathways are also better at degrading more recalcitrant compounds such as chlorinated organics.

6.2.3 Reactor Types

Most wastewater treatment processes are used for large volumes of wastewater with continuous flows, and are therefore continuous systems. However, for smaller and variable volumes, such as in sludge treatment and industrial wastewater treatment, fed-batch processes are used (*Process Science and Engineering for Water and Wastewater Treatment* – Unit 10). Their advantage is the lower capital cost of the plants required.

Wastewater reactors are also classified according to microbial habitats of the micro-organisms that utilise the wastewater. The design criteria include ones with attached growth and suspended microbial growth.

BIOLOGICAL REACTORS
Biological reactors can be either attached or suspended growth processes.

Suspended growth processes rely upon the microbes (Figure 6.1) being in free suspension within the wastewater. Contact with the wastewater constituents will reply upon good mixing in the reactor.

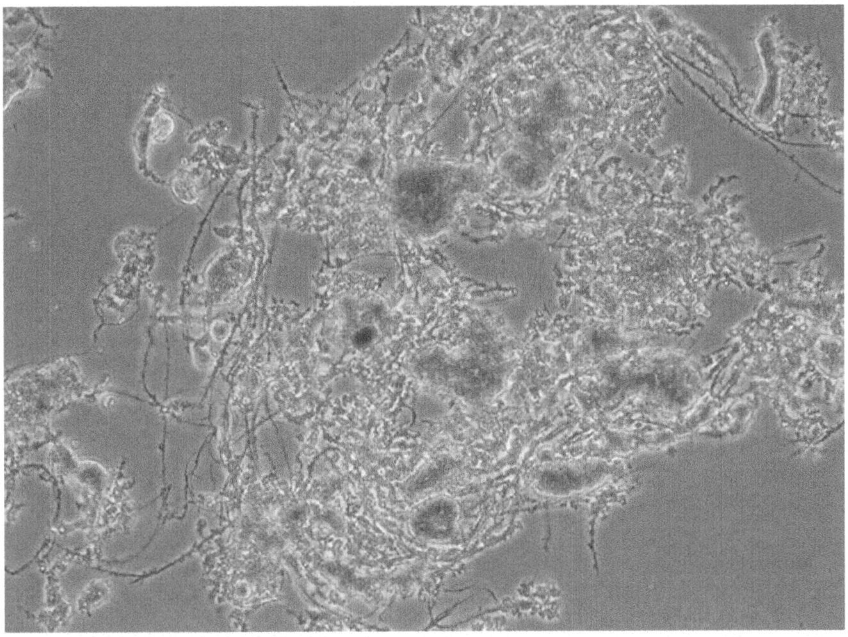

Fig. 6.1 – Suspended biomass

Fixed film processes provide an inert support material on which the micro-organisms can grow. The microbial population is then in contact with the wastewater constituents as it is passed over the support matrix.

In both types of systems, a separation stage such as sedimentation tank (Unit 3) is needed to remove the biomass from the clean effluent.

6.2.4 Treatment Systems

There are many types of biological processes that can be distinguished by the 3 categories described in Sections 6.2.2 and 6.2.3, i.e. oxygen requirements, flow type and microbial system. The most important of these are summarised in Table 6.4. It should be noted that there are many generic and proprietary processes on the market that can all be fitted into one or other of the categories listed.

BIOLOGICAL TREATMENT
All biological treatment processes require a stage to separate biomass from the effluent.

Table 6.4 – Types of biological wastewater treatment processes

Process	System	Reactor	Function	Configuration
Aerobic	Suspended-growth	Continuous	BOD removal	Activated sludge Oxidation ditch Contact stabilisation Aerobic pond treatment
Aerobic	Suspended-growth	Fed-batch	BOD removal Nitrification	Sequencing batch reactor
Aerobic	Fixed film	Continuous or fed-batch	BOD removal Nitrification	Trickling filtration High-rate filtration Rotating biological contractor Biological aerated filtration
Aerobic	Suspended-growth or Fixed film	Continuous	BOD removal Nitrification	Trickling filter solid-contractor process
Anoxic	Suspended-growth or fixed film	Continuous	BOD removal	Tertiary denitrification
Anaerobic	Suspended-growth	Fed-batch	Solids reduction	Sludge digestion
Anaerobic	Suspended-growth	Continuous	BOD removal Nitrification	Contact process Sludge blanket reactor Anaerobic pond treatment
Aerobic/ Anoxic	Suspended-growth or fixed film	Continuous	BOD removal Nitrification Denitrification	Multi-stage activated sludge filter
Aerobic/ Anoxic/ Anaerobic	Suspended-growth	Continuous	BOD removal Nitrification Denitrification Phosphate removal	Multi-stage activated sludge

6.3 Aerobic Fixed Film Processes

Section 6.3 requires a study time of about 2 hours

Fixed film processes provide an inert support material on which the micro-organisms can attach and grow over the surface forming a dense microbial biomass layer or *biofilm*. The microbial population is then in contact with the wastewater as it is passed over the support matrix on which the biofilm develops. However, not all fixed film reactors have a stationary support medium, in some reactor systems the medium itself move through the wastewater.

Fixed film aerobic processes are widely employed for BOD removal and denitrification in the treatment of:

- municipal sewage;
- food and beverage industry wastewaters;
- brewery and pharmaceutical wastewaters and
- oil refinery and textural wastewaters.

Fixed-film processes can achieve up to about 90% removal of BOD. The most widely used fixed-film reactors are trickling filter systems, other reactors include rotating biological contractors (RBC), biological aerated filters (BAFs), moving bed biofilm reactors and fluidised bed reactors.

6.3.1 Biofilm formation

Treatment in fixed film systems is dependent upon the activity of a biofilm which develops on a solid surface. Biofilms are common in nature because it is often more advantageous for a micro-organism in a flowing system to remain stationary and allow the flow of water to deliver nutrients rather than having to swim around in search of nutrients. A biofilm is composed of the cells of organisms, slimy material and sometimes particulate material that has become attached to the film. The organisms living in a biofilm may belong to a wide range of species of micro-organisms and macro-organisms, however, in the case of most wastewater treatment biofilms, bacteria form the basis of the biofilm. The biofilm organisms are held within a layer of slimy material. This slime is composed of complex polysaccharides and is produced by the biofilm bacteria. The slime is extremely important in the functioning of the biofilm as it plays an important role in adhesion, nutrient uptake and protection from anti-microbial chemicals.

BIOFILMS
A biofilms is a complex community of bacteria, fungi, protozoa and other macro-invertebrates that develop of surfaces which receive a constant supply of nutrients.

No inoculum is required to start biofilm growth. The organisms responsible for treatment come from the wastewater. Establishment may take some time, up to 3–4 weeks to achieve a reasonable level of BOD removal, longer for nitrification and around to 2 years to achieve the climax community.

Biofilm formation is a multi-stage process:

- Biofilm development is initiated by the adsorption of organic substances onto the filter media (the *conditioning layer*).
- The next stage is bacterial attachment to the conditioning layer.
- Bacteria continue to grow, and slime is produced, which helps to hold the biofilm together. Bacterial multiplication and slime production requires nutrient uptake. Biofilms are highly efficient at extracting nutrients from the passing solution but the exact mechanisms mediating this uptake are not well understood.

- The biofilm becomes colonised by other micro and macro-organisms.
- The *sloughing off* of the biofilm, which is caused in time by biofilm instabilities. This is a natural recycling and trimming of the biofilm.

The actual working of the biofilm is reliant on the transfer of nutrients and gases (e.g. oxygen in an aerobic system or carbon dioxide in an anaerobic system) through the biofilm layer, and waste products moving outwards to the layer of water (Figure 6.2).

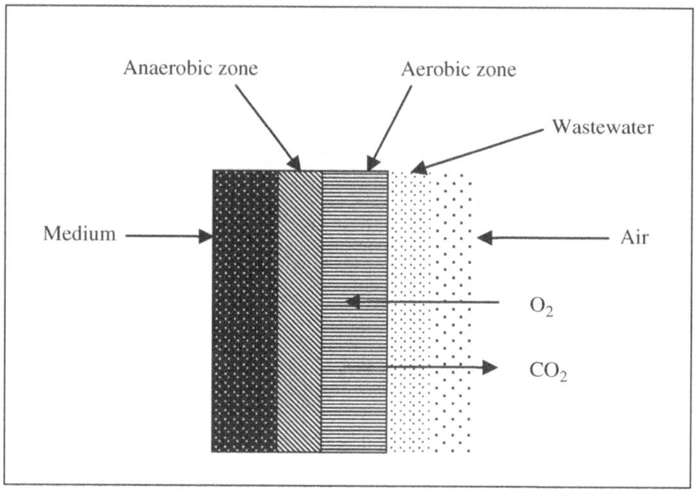

Fig. 6.2 – Biofilm structure (trickling filter)

As the bacteria on the solid/liquid interface metabolise the wastewater and reproduce, they gradually cause an increase in the depth of the slime layer. As the biological layer thickens, the bacteria in the interior layers find themselves in a nutrient limited situation, since the organic matter and oxygen (if aerobic) are utilised near the surface. Eventually, these interior cells die and lyse, breaking contact between the biofilm and the support medium. When sufficient cells have lysed, the biofilm will *slough off* and be carried from the filter by the waste flow.

6.3.2 Trickling Filters

Trickling or *percolating filtration* is the oldest biological process and in simplest form comprise of a bed of graded hard material (e.g. blast furnace slag) with a rotating distributor (Figure 6.3). In these systems, the settled wastewater is spread evenly over the surface of the filter bed (about 1.5 m in depth) by gravity feed distribution system, which can be used to regulate the volume and frequency of application of wastewater. The filter has a natural ventilating system that allows air to flows upwards through the interstices of medium ensuring all parts of the filter have sufficient oxygen (Figure 6.4). The treated effluent passes through the medium and is collected in a drainage system, the excess or sloughed biomass is removed by secondary settlement tanks downstream of the filter.

To improve the efficiency of trickling filter systems for strong organic wastewaters, blast furnace slag media can be replaced with light weight high surface area plastic media which allows for a much greater media depth (typically 4–5 m) and hence a smaller footprint for a given duty. Media types include: random (Figure 6.5) and modular media (Figure 6.6).

Fig. 6.3 – Trickling filters (courtesy of Anglian Water, UK)

TRICKLING FILTRATION
Wastewater is distributed over the surface a filter bed and percolates down through the medium while air is naturally ventilated up through the filter in order to supply sufficient oxygen for microbial growth.

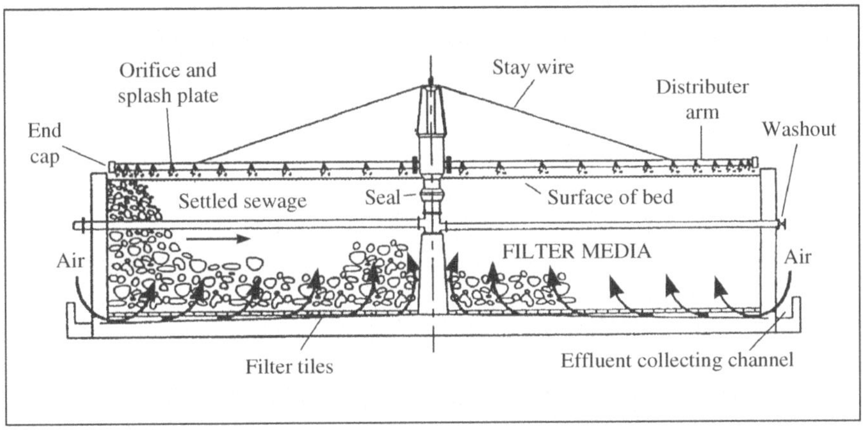

Fig. 6.4 – Trickling filter

TRICKLING FILTERS
Trickling filters can use random and structured (modular) media for their filter bed.

Fig. 6.5 – Random plastic filter bed media

Ex 6.2
What different configurations of trickling filters can be used to improved efficiency for industrial wastewaters?

There are also a number of modification to a conventional single-pass filtration that can be used improve it removal efficiency. These include the use of *roughing* or *high-rate* filter prior to a low-rate secondary filter, known as two-stage filtration, *recirculation* which dilutes the incoming wastewater with returned final effluent and *alternating double filtration* (ADF) that employs two

Fig. 6.6 – Modular filter bed media

identical filter in series using identical filter media, where the sequence of the filters are reversed every 1–2 weeks to prevent clogging of the first filter.

Trickling filter systems are generally reliable and problem free. However, clogging of the filter medium can cause *ponding* on the filter bed surface and flies are able to breed in the biofilm causing a nuisance.

6.3.3 Rotating Biological Contactors

Rotating biological contactors (RBC) consist of series of flat or corrugated discs and the biomass grows on the surface of the discs which rotate on a central shaft and are half-submerged in the reaction tank. The water to be treated flows through the tank parallel to the shaft and over the submerged surfaces of the discs. Aeration is achieved by the rotation of the discs so that half of the disc is always above the liquid level and exposed to air. Some RBC design incorporate primary settlement, but all allow for secondary settlement. RBC units are usually covered to protect the medium and biofilm from the weather (Figure 6.7). The covers also reduce noise, eliminate fly nuisance and control odour to some extent.

Fig. 6.7 – Rotating biological contactor (courtesy of Severn Trent Water, UK)

6.3.4 Biological Aerated Filters

Biological aerated filters (BAF) are submerged, fixed-media, three phase reactors for wastewater treatment. The main feature of a BAF is the combination of bioreactor with subsequent biomass separation in one reactor. The granular media of a BAF has two roles. It acts as support for biomass that will undertake carbonaceous BOD, nitrogen and even phosphate removal when operated under alternating anaerobic/aerobic conditions. Secondly, the media has a depth filtration action (Unit 3) preventing biomass leaving with the effluent stream; however this results in little natural sloughing so the regular backwashing is required to prevent the medium becoming blocked. Flow in a BAF can be operated in an upwards or downward direction with air being supplied by diffusers at the base of the reactor (Figure 6.8).

BAFs are a small footprint alternative to trickling filter and activated sludge processes. There is a great deal of flexibility in the use of BAFs; such reactors have been used for secondary treatment and also tertiary treatment of municipal wastewaters, in combination with nutrient removal and for industrial effluents.

BIOLOGICAL AERATED FILTERS
BAFs are submerged fixed-film reactors that combined biological degradation with biomass separation.

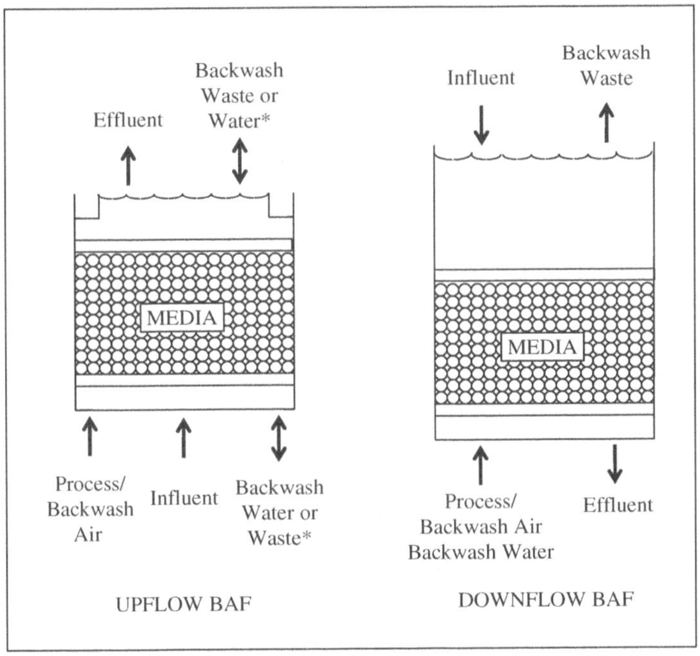

Fig. 6.8 – Biological aerated filters

6.3.5 Fluidised Beds

Fluidised beds contain media that is in constant motion rather than static as in BAFs. This means that the media particles are not in contact with each other so the whole surface area is available for biomass support. These reactors can thus accept higher organic loading rates than any other type of aerobic reactor. A high upflow liquid velocity is required in order to fluidise the particles to provide the large available surface area. A high density of biomass (concentrations up to 40,000 mgl^{-1}) is maintained without the need for recirculation, which means that an oxygen supply is required since oxygen limitation would be a problem if air was used. Oxygen is injected into the influent stream as it enters the base of the reactor (Figure 6.9). The expansion of the media is controlled by the rate of wastewater input, thus ensuring that the effluent is removed without a loss of the media or solids. Fluidised bed systems are highly compact, but expensive to run due to the pure oxygen and the greater use of pumps.

FLUIDISED BEDS
Fluidised beds are high-rate bioreactors with inert media in constant motion.

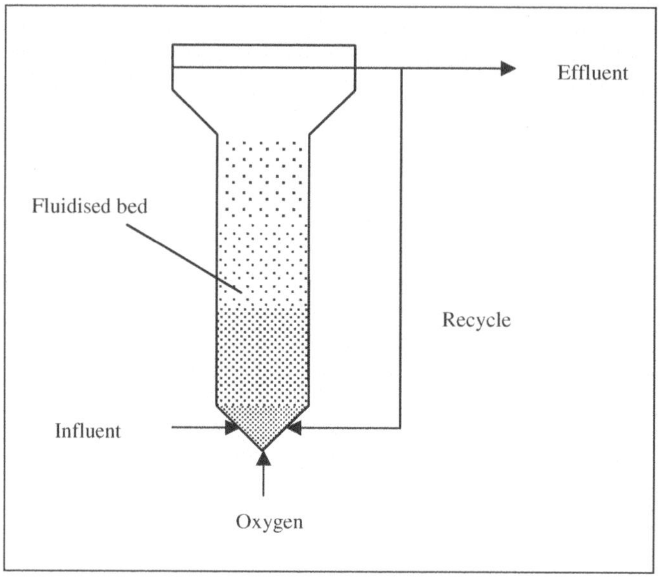

Fig. 6.9 – Fluidised bed reactor

6.4 Aerobic Suspended Growth Processes

Section 6.4 requires a study time of about 2 hours

In suspended growth processes the micro-organisms are either free-living or flocculated to form suspended biomass or flocs (Figure 6.1) which contain a variety of bacteria, fungi and protozoa. These flocs are mixed with wastewater in an *aeration tank* by aerators that not only supply the oxygen, but also maintain the biomass in suspension to ensure contact between the nutrients in the wastewater and microbial population. Aeration tanks may be either completely mixed or plug flow reactors (*Process Science and Engineering for Water and Wastewater Treatment* – Unit 10).

Suspended growth processes are primarily used to treat domestic wastewater, but are also used in the treatment of wastewater from a wide range of industries including:

- food processing;
- brewing;
- soft drinks manufacture;
- pharmaceuticals;
- paper making and
- oil refining.

FLOCS
Suspended biomass or flocs are dense microbial population of bacteria, fungi and protozoa that utilise organic matter as a food source and interact to maintain a dynamic population.

Suspended growth processes typically provide up to 90% removal of BOD and are generally less susceptible to upset than fixed – film processes. The most widely used suspended growth reactors are activated sludge systems (Figure 6.10); other reactor designs include sequencing batch reactors (SBR), biological nutrient removal (BNR) systems and membrane bioreactors (MBRs).

Fig. 6.10 – Activated sludge aeration lane (courtesy of Severn Trent Water, UK)

6.4.1 Activated Sludge

The activated sludge process relies on a dense population of micro-organisms being in mixed suspension with the wastewater (the *mixed liquor suspended solids*) under aerobic conditions. This is achieved either by blowing the air under pressure into diffusers which create bubbles or by entraining air from above the surface by a turbine mixing device called a *surface aerator* (Figure 6.11). Both systems use a considerable amount of energy.

ACTIVATED SLUDGE
Control parameters in activated sludge systems are mixed liquor suspended solids; sludge age; *F:M* ratio and sludge settleability.

Fig. 6.11 – Activated sludge surface aeration (courtesy of Anglian Water, UK)

The *mixed liquor* flows from the aeration tank into a settling tank where the sludge is separated and returned to the reactor as *returned activated sludge*, and the clarified *final effluent* is discharged (Figure 6.12). The time that the microbial sludge spends in the activated sludge process (both within the aeration tank and while being separated and returned) is known as the *solids residence time* (SRT) or *sludge age*. The amount of suspended solids in the treatment process is regulated through the recycling of the activated sludge and periodically discharging of the excess sludge from the settling tank as *surplus* or *waste activated sludge*. The relationship between the organic matter (substrate concentration) and the sludge biomass (microbial concentration) is a fundamental in maintaining maximum organic matter removal and is referred to as the food to micro-organisms (*F:M*) ratio. Conventional activated sludge have a *F:M* ratio of between 0.2–0.4. The sludge settleability determined by *sludge volume index* (SVI) is another parameter used to ensure good separation in secondary settlement, so that the final effluent has a low suspended solids concentration and sufficient biomass is returned to the aeration tank to maintain microbial activity.

Activated sludge processes are particularly versatile and are well suited to modification. Examples include:

- the use of pure oxygen instead of air, results in increased oxygen transfer and increases the potential rate of BOD removal;
- the use of tapered aeration to supply oxygen according to respiratory requirements of the biomass;
- incremental feeding in order to proportion the BOD loading with the oxygen demand and oxygen utilisation rate and
- inclusion of anoxic zones for denitrification (Section 6.3.3).

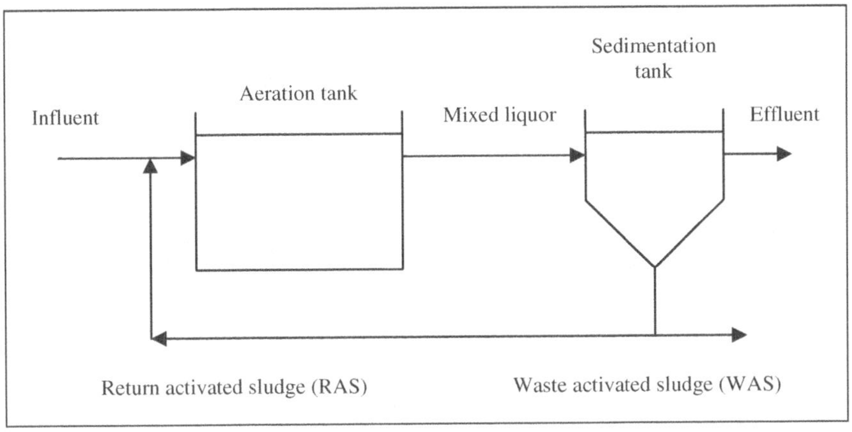

Fig. 6.12 – Conventional activated sludge process

6.4.2 Sequencing Batch Reactors

The *sequencing batch reactor* (SBR) is a fed-batch suspended growth reactor with mixing, aeration and decanting in the same tank. Wastewater is filled into the tank either with or without aeration, aerated for a pre-set time, allowed to settle under quiescent conditions and the supernatant decanted (Figure 6.13). The sludge remains in the reactor until reduction is required. SBRs are very flexible in that the sequence and timings of the operating cycle can be easily changed to suit variations in influent quality, which generally requires a more complex control process, but the process does offer a reduced footprint because there is no final settling tank.

SEQUENCING BATCH REACTORS
SBRs are fed-batch suspended growth processes where aeration and biomass separation occur in the same tank.

6.4.3 Biological Nutrient Removal

In response to stricter discharge standards for nutrients from wastewater treatment plants, biological nutrient removal (BNR) has become increasingly common and many different design configurations based on activated sludge are now widely used. These processes incorporate nitrification, denitrification, phosphorus removal and BOD removal by providing a series of aerobic, anoxic and anaerobic zones.

Conventional activated sludge processes can be adapted for denitrification by addition of an anoxic zone (Figure 6.14). In conventional activated sludge processes (Figure 6.12), nitrification can be achieved by increasing the sludge age or biomass residence time. This encourages the growth of the bacteria that convert ammonia to nitrate (Table 6.1). An internal recycle ensures that nitrate is returned to the anoxic zone for denitrification using the BOD from the influent.

Phosphorus removal during wastewater treatment can also be enhanced by its biological uptake. This can be achieved by subjecting the mixed liquor to a cycle of anaerobic and aerobic conditions. The simplest biological phosphorus removal system is the anaerobic–oxic (A–O) process for non-nitrifying activated sludge (Figure 6.15). In the anaerobic tank bacteria release phosphorus that is then absorbed to excess by the bacterial biomass in the aeration tank.

Ex 6.3
What are the 5 cycles in a SBR plant?

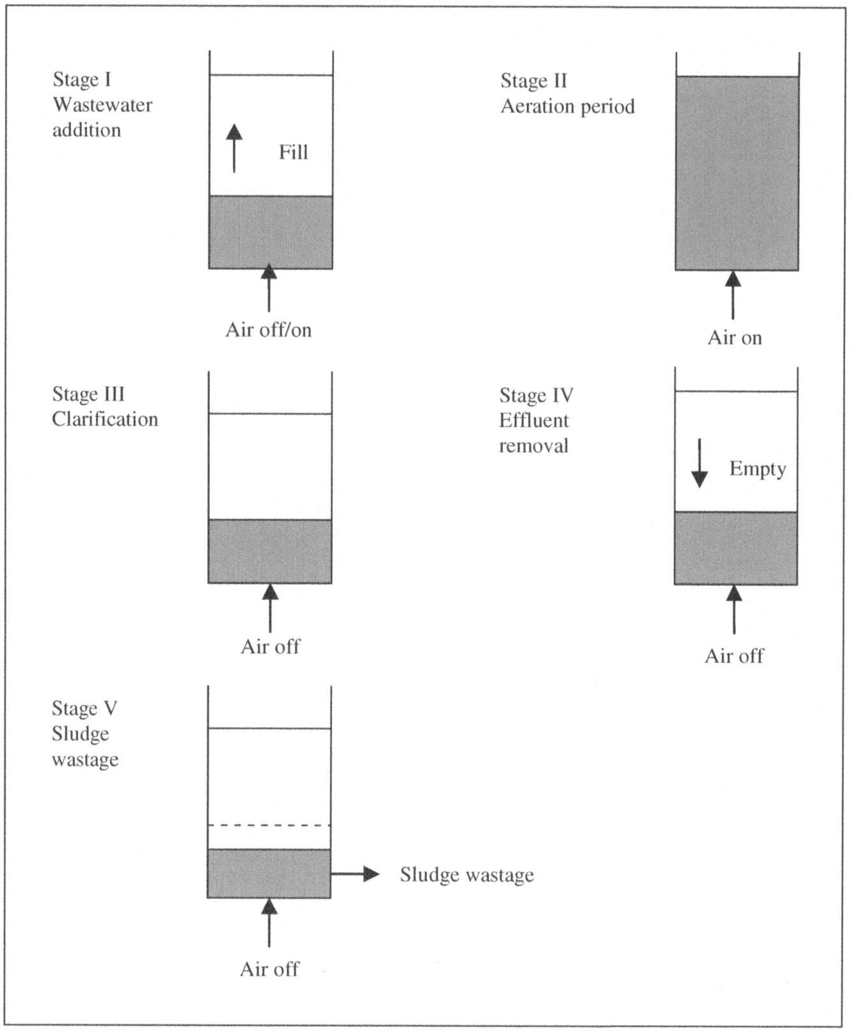

Fig. 6.13 – Typical operating cycles of an SBR

Ex 6.4
What series of aerobic, anoxic and anaerobic zones in an activated sludge plant can be used to improve?
a) denitrification
b) phosphorus removal

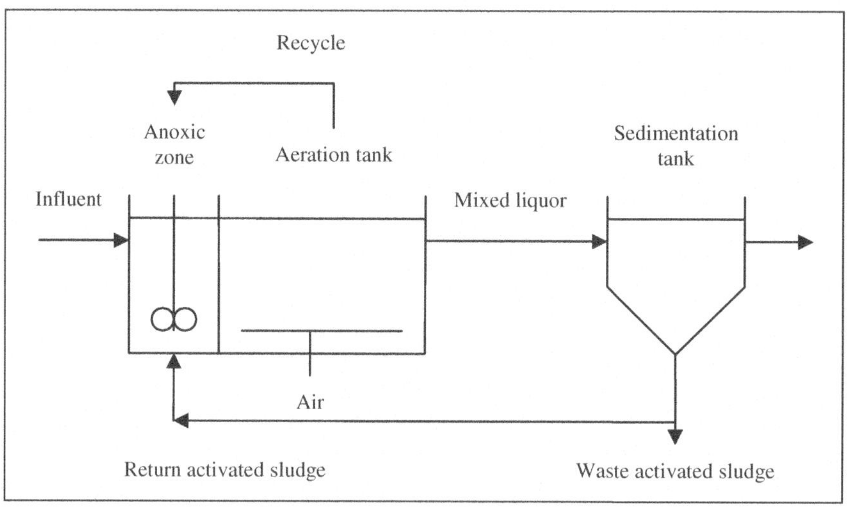

Fig. 6.14 – Pre-denitrification activated sludge

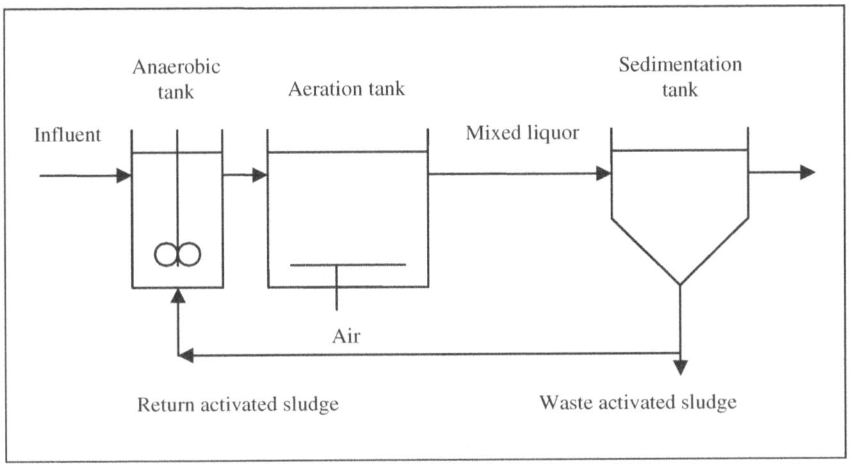

Fig. 6.15 – Two-stage (A–O) biological phosphorus removal process

More advanced BNR systems include the Bardenpho, University of Cape Town (UCT), Virginia and the AAO (anaerobic, anoxic and oxic) processes, all of which incorporate nitrification, denitrification and phosphorus removal in various sequences. SBRs can also be configured to have a process cycle that includes an anaerobic, aerobic and anoxic stage for BNR. All these processes are between 3 and 5 stages with a combination of anoxic, anaerobic and aerobic stages.

6.4.4 Membrane Bioreactors

Membrane bioreactors (MBR) are the amalgamation of a suspended growth reactor and membrane filtration device into a single unit process. The membrane unit can be configured external to, as in side stream operation (Figure 6.16) or immersed in the bioreactor (Figure 6.17). The main difference between the two configurations is in their operation. In the case of an external system the membrane is independent of the bioreactor. Feed enters the bioreactor where it contacts biomass. This mixture is then pumped around a recirculation loop containing a membrane unit where the permeate is discharged and the retentate returned to the tank. The transmembrane pressure and the crossflow velocity, which define the operation of the membrane, are both generated from a pump (Unit 7).

Immersed systems differ in that there is no recirculation loop as the separation occurs within the bioreactor itself. Under these circumstances the transmembrane pressure is derived from the hydraulic head of the water above the membrane. In some systems this is supplemented by a suction pump to increase the transmembrane pressure, although this remains significantly less than in side stream operation. Fouling control is achieved by scour at the membrane surface. The energy is derived from the aeration process with the movement of bubbles close to the membrane surface generating the necessary liquid shear velocity.

MEMBRANE BIOREACTORS
MBR can be configured with either as a side stream or an immersed membrane filter in order to separate the permeate from the mixed liquor.

The coupling of a membrane to a bioreactor offers inherent advantages over conventional biological wastewater treatment systems; these include process intensification and product water quality. The permeate from the membrane is free from solids and macro-collodial material, typical water product qualities

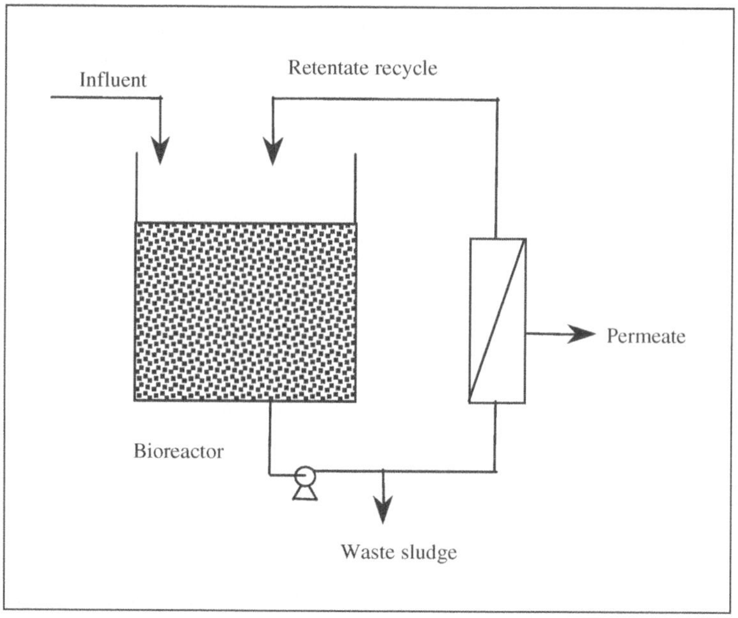

Fig. 6.16 – MBR with side stream membrane

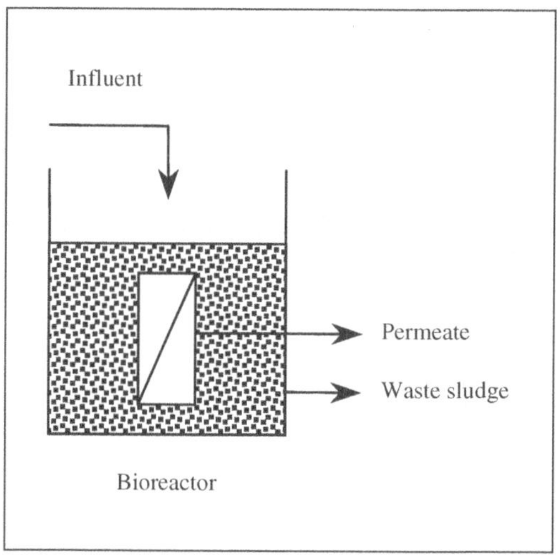

Fig. 6.17 – MBR with immersed membrane

are <5 mgl^{-1} suspended solids, often down to 1–2 mgl^{-1} and <1 NTU turbidity. Complete retention of all suspended matter is attainable, including bacteria and viruses, such that MBR effluents can be of a quality suitable for discharge to sensitive regions.

Sludge age and hydraulic retention time are completely independent. Thus, MBRs can be operated at low HRTs and long SRT's without washout of biomass in the effluent which is common problem in activated sludge. The membrane effectively nullifies problems such as filamentous growth and sludge degassing.

6.5 Anaerobic Processes

Section 6.5 requires a study time of about 2 hours

Anaerobic processes are treatment systems in which the heterotrophic micro-organisms are not in contact with dissolved oxygen or the alternative electron acceptor, nitrate. The bacteria use carbon substrates as the electron acceptor (*Process Science and Engineering for Water and Wastewater Treatment* – Unit 4). However, such biochemical routes for obtaining energy are less efficient than aerobic systems, so anaerobic micro-organisms tend to be slower growing. This is important when considering how and where anaerobic systems can be used. Anaerobic processes are used to treat strong organic wastewater (BOD $> 500\ mgl^{-1}$) and for further treatment of primary and secondary sludges from conventional wastewater treatment. Therefore, since anaerobic bacteria are slow growing; larger reactor volumes are generally required. The advantages and disadvantages of anaerobic treatment are outlined in Table 6.5.

ANAEROBIC PROCESSES
Anaerobic microbes have slower growth rates than aerobic microbes; therefore anaerobic processes are generally larger and have long retention times.

Table 6.5 – Advantages and disadvantages of anaerobic treatment

Advantages	Disadvantages
No aeration required	Slow to start-up
Low sludge production	Long retention times
Sludge is highly stabilised	Sensitive to toxic compounds
Low operating costs	High capital costs
Methane produced as end-product	Production of odorous and corrosive compounds
Can treat variable loads	Requires post-treatment to remove BOD
	Can require heating

6.5.1 Microbiology and Biochemistry

Anaerobic treatment, like aerobic processes, is not undertaken by one type of bacterium, but a whole “ecology”. However, it should be noted that anaerobic systems are almost exclusively based on bacterial communities, with some fungi present.

ANAEROBIC TREATMENT
Anaerobic treatment consists of 3 phases:
i) hydrolysis,
ii) acetogenesis and
iii) methanogenesis.

Anaerobic treatment can be considered to occur in three stages. Firstly, complex biochemicals such as proteins, lipids, carbohydrates and nucleic acids are broken down to long and short chain fatty acids, such as propionate and butyrate (*Process Science and Engineering for Water and Wastewater Treatment* – Unit 5). This is called *hydrolysis* or the *fermentative stage*. The second stage involves the conversion of these molecules to acetic acid. This is termed *acetogenesis*. The third stage involves the conversion of acetate, and also hydrogen produced by *obligate hydrogen producing acetogens* (OHPA), into methane. This is termed *methanogenesis*.

In the initial breakdown of the complex biochemicals during hydrolysis the following conversions occur:

- *proteins to amino acids*
- *lipids to long chain fatty acids and alcohols*
- *carbohydrates to monosaccharides (sugars)*
- *nucleic acids to sugars*

These reactions are carried out in treatment processes extracellularly i.e. outside the micro-organism cell wall. Enzymes that undertake the hydrolytic reactions, such as lipases, proteases and cellulases, are secreted by the bacteria. Different bacteria are able to attack the different biochemicals. Once broken down, the smaller compounds are taken back into the cell, where they are further broken down into long and short chain fatty acids (acetogenesis). The bacteria responsible for producing these fatty acids are fast growing which can result in an excess of these fatty acids. This can cause the pH to drop, which can lead to souring of an anaerobic system.

Clostridium and *Micrococcus* are the bacteria mainly responsible for lipid hydrolysis and can number 10^4–10^5 per ml. *Clostridium* (e.g., *C. butyricum*, *C. perfringens*) are also responsible for *proteolytic activity* and can number up to 10^6 bacteria per ml. Proteolytic enzymes have a broad pH spectrum 5.0–11.0. Degradation routes for cellulose and lignin polysaccharides are not well understood in anaerobic treatment. Amylases for degrading starch and related polysaccharides are produced by *Bacillus* species among others.

Acetogensis can be considered as having two sub-stages and has some overlap with the acidogenic stage. Firstly, the breakdown of the products of acidogenesis (long chain fatty acids etc.) to short chain fatty acids such as propionate, butyrate, valerate and iso-valerate, in addition to some acetate. The second sub-stage is the breakdown of these short chain fatty acids to acetate. Some of this acetate is produced concomitantly with hydrogen by OHPA bacteria.

There are two sub-groups of methanogens: those like *Methanothrix (Methanosaeta), Methanosarcina* and *Methanospirillum* that are able to produce methane direct from acetate, and those like *Methanobacterium, Methanobrevibacterium* and *Methanoplanus* that produce the gas directly from hydrogen and carbon dioxide.

The overall reaction in anaerobic treatment can be summarised as follows:

$$\mathbf{C_6H_{12}O_6 + 2H_2O \Rightarrow 2CH_3COOH + 4H_2 + CO_2}$$

$$\mathbf{CO_2 + 4H_2 \Rightarrow CH_4 + 2H_2O}$$

$$\mathbf{CH_3COOH \Rightarrow CH_4 + CO_2}$$

6.5.2 Anaerobic Systems

Anaerobic systems can be run at different temperatures. Generally, the temperature is around 35°C, i.e. mesophilic conditions. However, thermophilic operation at 55°C is also possible.

THERMOPHILIC PROCESSES
Thermophilic anaerobic processes offer high wastewater degradation than mesophilic systems due to increased chemical and microbial growth rates at 55°C.

Thermophilic operation of anaerobic reactors has been shown to be a feasible alternative to mesophilic temperatures. Examples such as the anaerobic treatment of palm oil mill effluent, pharmaceutical wastes, distillery wastes and ice-cream wastes have all been successful. Thermophilic systems are capable of high reaction and growth rates, enabling a rapid biodegradation of wastes, with an increased reduction in the viability of any pathogenic organisms due to the elevated temperatures. Although thermophilic systems require more energy to maintain the higher temperature, this can be supplied by the biogas produced. Another alternative would be to treat wastewaters that are released from the process at high temperatures, where cooling would be required prior to mesophilic treatment.

Thermophilic operation has a number of drawbacks. Thermophilic reactors tend to be less reliable, with a reduced COD or BOD removal, gas production and methane yield. Stability of the sludge may also be a problem. The presence of inhibitory and toxic components of the wastewater has been known to cause more problems under thermophilic conditions than with mesophilic conditions.

6.5.3 Anaerobic Reactors

Both fixed-film and suspended growth reactors have been used for anaerobic treatment. The most common are anaerobic sludge digesters, which are continuously stirred tank reactors (CSTR) and are used primarily in the treatment of sewage sludge (Unit 8). Most other types of anaerobic reactors are used for the treatment of wastewater. These include anaerobic contact reactors, anaerobic filters, anaerobic fluidised bed reactors (AFBR) and upflow anaerobic sludge blanket reactors (UASB). Typical COD removals from these reactors are in the range of 70–80%.

ANAEROBIC REACTORS
Anaerobic reactors most provide separation for three end products:
i) sludge
ii) waste liquor and
iii) gas.

Anaerobic contact reactors (Figure 6.18) are similar in principle to that of activated sludge. The active anaerobic biomass is separated from the final effluent by sludge settlement or other separation processes (such as membrane filtration or centrifugation) and is returned to the reactor to maintain a high biomass content. The major operational problem is poor settlement due to the attachment of gas bubbles.

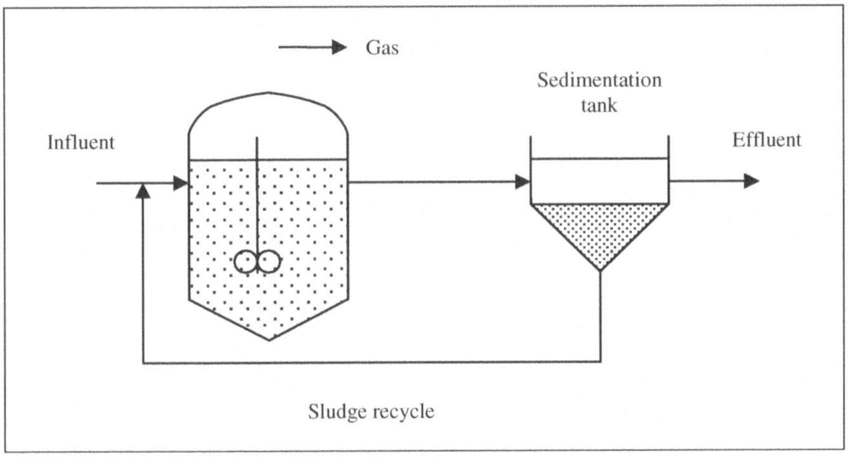

Fig. 6.18 – Anaerobic contact reactor

Ex 6.5
What would a decrease in gas production indicate in anaerobic treatment?

Anaerobic filters are similar in principle to trickling filters in that a static media is used to retain the active biomass and that the filters are completely flooded so that no oxygen can enter the reactor. They are operated in either an up-flow or down-flow mode with a wide variety of media from mineral to random and structured plastics. At present the major use of anaerobic filter is as a denitrification stage.

Anaerobic fluidised bed reactors (AFBR), like aerobic fluidised bed reactors, consist of fluidised media that are coated with biomass. The medium is remains in suspension by maintaining a high rate of gas and effluent recycle (Figure 6.19). Although anaerobic fluidised beds are difficult to start up, they are highly resistant to temperature, toxic compounds and high organic loadings.

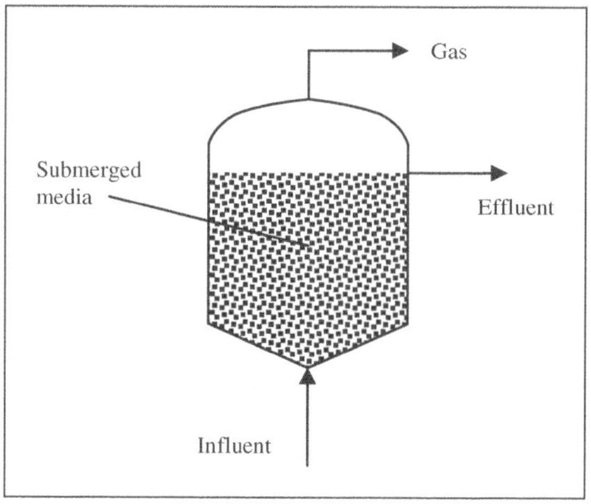

Fig. 6.19 – Anaerobic filter

Upflow anaerobic sludge blanket reactors (Figure 6.20) consist of a biological tank with the upflow combined with a settling stage (Figure 6.21) and rely on the propensity of anaerobic biomass to aggregate into dense flocs or granules. By controlling the feed of the wastewater through the base of the reactor, the granules remain in the tank as a *sludge blanket* as their settling velocity is greater than the upflow water velocity, thus preventing washout of the biomass. Compared with other anaerobic contact processes, UASB are difficult to operate and to maintain the structure of the sludge blanket; also a suitable inoculum is sometimes required to start-up the process.

Fig. 6.20 – Upflow anaerobic sludge blanket reactor (courtesy of Wageningen University, The Netherlands)

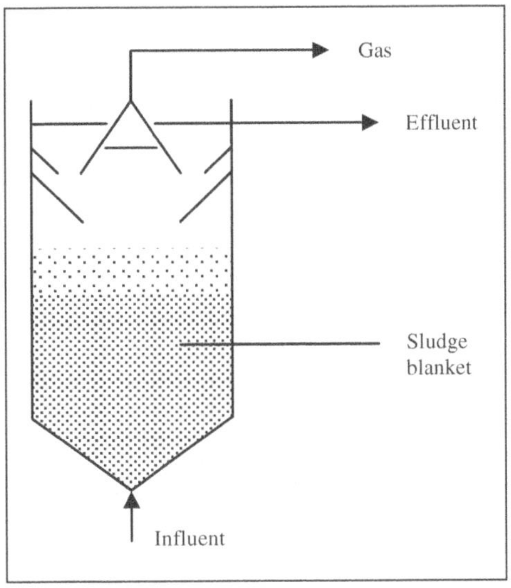

Fig. 6.21 – Upflow anaerobic sludge blanket reactor

Typical operating conditions for the various anaerobic reactor types are summarised in Table 6.6.

Table 6.6 – Operating parameters for anaerobic reactors

Reactor Type	HRT (d)	SRT (d)	SS	COD	Loading (kg COD.m^{-3}.d^{-1})
CSTR	10–60	10–60	high	low–high	0.25–3
Contact reactor	12–15	20	low	low	0.25–4
Anaerobic filter	0.5–12	20	high	low–high	1–40
AFBR	0.2–5	30	low	low–v.high	1–100
UASB	0.5–7	20	low	low–high	10–30

6.6 Self Assessment Questions

SAQ 6.1 Define the two reactor systems that are used by microbes to grow in biological wastewater treatment processes.

Describe the different growth mechanisms that are employed by microbes (biomass).

SAQ 6.2 What are MBRs and describe the principle configurations that are used in wastewater treatment?

SAQ 6.3 Describe the different types of mechanism microbes use in wastewater based on utilisation of oxygen?

Discuss the advantages and disadvantages of these two approaches and give 2 specific examples.

SAQ 6.4 Activated sludge is one of the most common wastewater treatment processes, describe the principle stages and essential components in this suspended growth system.

SAQ 6.5 What are the principle microbiological and biochemical phases in an anaerobic treatment system?

Describe the different reactor types used in the anaerobic treatment of wastewater.

SAQ 6.6 Discuss briefly the mechanisms involved in the formation of biofilms.

Give three examples of the use of biofilms in biological processes and describe the operation briefly.

6.7 Glossary of Terms

Dry Matter (DM) *or Total Solids* (TS)
This is a measurement of the solids present in a known volume of sample. All the water is removed from the sample by drying at 105°C.

Suspended Solids (SS)
A known volume of sample is filtered or centrifuged. The solids are then dried.

Volatile Suspended Solids (VSS)
The suspended solids are dried and then ashed at 500 to 600°C. This removes all the organic matter, leaving the inorganic, or ash.
The volatile suspended solids can be calculated by:

$$\mathbf{VSS = SS - ash}.$$

The organic matter assumed to be the sludge.

Mixed Liquor (ML)
This is the mixture of sludge and solution in the aeration basin of the activated sludge tank.

Mixed Liquor Suspended Solids (MLSS)
The amount of sludge in the ML. More often it is described as MLVSS (volatile solids), that is the active biomass.

MLSS is approximately 4kg dry weight/m^3 (g/dm^3) or 4000 ppm.
Then the dry weight of cells/m^3 (x) $=$ MLVSS ≈ 3

Typically, 65–75% of SS is VSS and in VSS viability approximately 50%.

Sludge Volume Index (SVI)
The SVI is especially important with activated sludge as it can indicate problems with the sludge.

It is measured by the volume (cm^3) occupied by 1 g of sludge.
A good activated sludge has an SVI between 40–60 cm^3/g, while a bulking sludge has an SVI of >200 and a poor sludge, >120.

Sludge Loading Rate (SLR) *or Food:Microbes Ratio* (F/M)
This can be thought of as the removal of BOD per unit biomass per day.
In the case of activated sludge, a unit needs to be able to handle 3 times the dry weather flow.

$$\mathbf{SLR = \frac{kgBOD/day}{kg\,MLSS} = \frac{FS_R}{Vx}}$$

Typically SLR $= 0.15$kg BOD/kg MLSS/d

Volumetric Loading Rate (VLR)
This is defined as the substrate added per unit volume per day.

Hydraulic Retention Time (HRT)
This is also called the retention time or the detention time, and will vary according to each type of biological process. For example, with activated sludge, the HRT is at least 5 hours.

$$\mathbf{HRT = V/F}$$

Sludge Retention Time (SRT)
This may also be called the solids (cells) mean residence time (SRT, θ_x) or, especially with activated sludges, the sludge age.

$$\mathbf{SRT} = \frac{\textbf{mass of cells in reactor}}{\textbf{mass flow rate leaving}}$$

The SRT can vary according to the process. For activated sludge it is in the range 10–20 days.

Floc Loading
This gives an indication of the relative amounts of BOD and solids in the system and will vary along plug flow units.

6.8 Solutions to Exercises

Ex 6.1 Biological wastewater treatment system utilises naturally occurring micro-organisms from the wastewater. These organisms utilise the soluble and colloidal organic and inorganic matter as an energy and carbon source.

Ex 6.2 Single-stage filtration can be improved by the use of a roughing or high-rate filter prior to a low-rate secondary filter (two-stage filtration), the incursion of recirculation which dilutes the incoming wastewater with returned final effluent and alternating double filtration which use two identical filter in series who's sequence in changed every 1–2 weeks to reduce media blocking.

Ex 6.3 SBR cycles: stage I – wastewater addition; stage II – aeration period; stage III – sediment; stage IV – effluent removal and stage V (optional) – Sludge wastage.

Ex 6.4

a) Denitrification can be improved by the addition of anoxic zone prior to aeration tank, known as conventional pre-nitrification
b) Phosphorus removal can be improved biological by the addition of anaerobic zone prior to the aeration tank. The anaerobic zone solublises all the phosphorus, which is adsorbed by the microbial biomass during aeration, referred to as an A–O process.

Other more complex BNR can also employed that incorporate nitrification, denitrification and phosphorus removal in a series of anaerobic, anoxic and aerobic zones.

Ex 6.5 A decrease in gas production would be the first indication of process is unstable, methanogens are being inhibited.

Bibliography

Unit 6 – Biological Processes

Bitton G. *Wastewater Microbiology*. 2nd ed., John Wiley and Sons, Inc. New York (1999).

Gray N.F. *Biology of Wastewater Treatment*. 2nd ed., Imperial College Press, London (2004).

Henze M., Harremoes P., la Cour Jansen J. and Arvin E. *Wastewater Treatment: Biological and Chemical Processes*. 3rd ed., Springer-Verlag, Berlin (2002).

Janssen P.M.J., Meinema H.F. and van der Roest H.F. *Biological Phosphorus Removal*. IWA Publishing, London (2002).

Rittmann B.E. and McCarty P.L. *Environmental Biotechnology: Principles and Applications*. McGraw-Hill, New York (2001).

Seviour R.J. and Blackall L.L. *The Microbiology of Activated Sludge*. Kluwer Academic Publishers, Dordrecht (1998).

Speece R.E. *Anaerobic Biotechnology for Industrial Wastewaters*. Archae Press, Tennessee (1996).

Wilderer P.A., Irvine R.L. and Goronszy M.C. *Sequencing Batch Reactor Technology*. IWA Scientific and Technical Report Series No. 10, IWA Publishing, London (2001).

Wuertz S., Bishop P.L. and Wilderer P.A. *Biofilms in Wastewater Treatment*. IWA Publishing, London (2003).

7. Membrane Processes

Contents

Aims and Objectives

This Unit covers the principles of membrane processes used in water and wastewater treatment works:

After studying these notes you should be able to:

1. explain in your own words the following terms and concepts:

   membrane separation
   microfiltration
   ultrafiltration
   reverse osmosis
   electrodialysis

2. describe the key parameters in membrane process performance.

It is important that you are able to complete all the self assessment questions at the end of this Unit.

Essential Prerequisites

It is not necessary to have a completed any other Units before undertaking this Unit.

Additional, information on the fundamentals for this Unit, refer to the following Units in *Process Science and Engineering for Water and Wastewater Treatment*:

Unit 1	Fundamentals of Water Chemistry
Unit 7	Fundamentals of Process Engineering
Unit 11	Engineering Hydraulics

7.1 Introduction

Membrane processes were first used in water treatment in the 1960s and are now an established technology. Although essentially a physical separation process, the range of membrane types and applications is such that *membrane separation* is generally regarded as a separate technology.

In this Unit the principles involved in the manufacture and classification of different membrane configurations and their application in water and wastewater treatment are outlined.

Section 7.2 requires a study time of about 1/2 hour

7.2 Membranes and Membrane Processes

There are many different definitions of *membrane* which vary considerably in comprehensiveness and clarity. We can define a membrane as:

> *A material through which one type of substance can pass more readily than others, thus presenting the basis of a separation process*

This definition identifies the property of the membrane to separate components of the water to be treated, and it is this property that is of key interest when selecting or designing membrane separation systems for water treatment. Membrane processes can achieve selective removal of contaminants across a wide range of sizes (Figure 7.1), with ultimate selectivity for ions over water.

MEMBRANES
Membranes achieve selective removal of contaminants based on size separation.

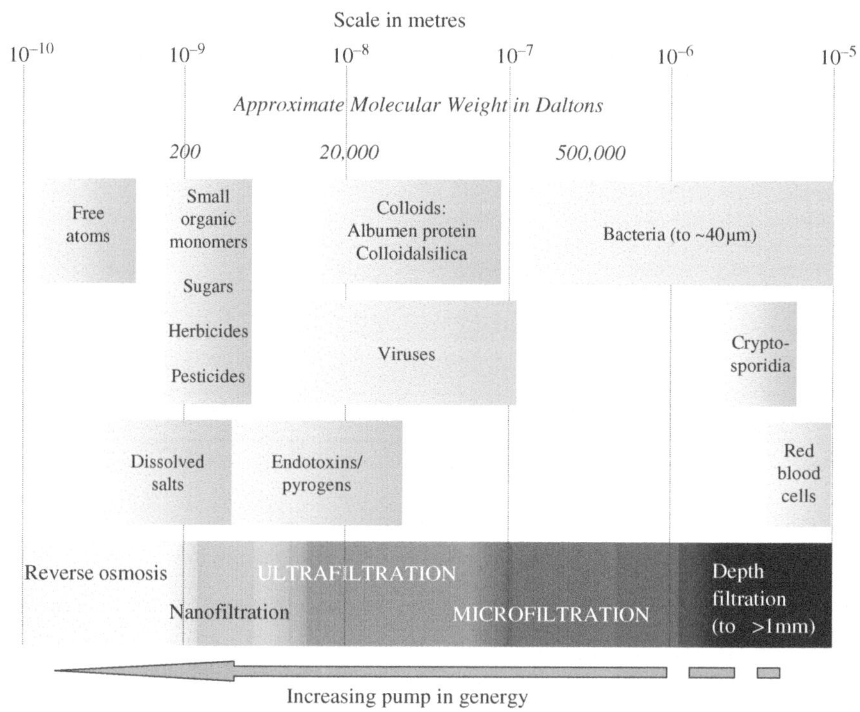

Fig. 7.1 – Membrane processes

Ex 7.1
Identify which of the following can be removed by ultrafiltration:
a) hardness ions
b) alkalinity
c) pyrogens
d) ethanol
e) colloidal proteins

7.3 Membrane Structure

Section 7.2 requires a study time of about 1 hour

The principal objective in membrane manufacture is to produce a material of reasonable mechanical strength, and which can maintain a high flux of a desired permeate with a high degree of selectivity. These last two parameters are mutually counteractive, but the optimum condition is always a thin layer of material with a narrow range of pore size.

The range of available membrane materials is very diverse as Table 7.1 shows. Materials vary widely in chemical composition and physical structure, but the most fundamentally important property is the mechanism by which separation is actually achieved. On this basis alone, all membranes may be placed in one of two categories: *porous* and *dense*.

POROUS MEMBRANES
Microfiltration, ultrafiltration and nanofiltration are common porous membranes processes that achieve removal of contaminated based on mechanical separation.

Porous membranes achieve separation mechanically, i.e. by sieving or straining. Processes which operate by this mechanism are termed *filtration* processes and include *microfiltration* (MF), *ultrafiltration* (UF) and *nanofiltration* (NF). These processes are relatively low in energy demand, since the pores are large compared with those of dense membranes.

Dense membranes have chemical selectivity, in other words they are able to achieve separation of components through physico-chemical interaction between the permeating components and the membrane material. They are the membrane processes having the highest selectivity and thus the highest energy demand. In *reverse osmosis* (RO) (sometimes called *hyperfiltration* in the US) and nanofiltration, separation of dissolved ions is achieved by virtue of differing *solubility* and *diffusion* rates of water (*solvent*) and dissolved species (*solutes*) in water. In *electrodialysis* it is achieved by virtue of differing ionic charge and charge density of solute ions, using ion-exchange membranes with allow selective passage of ions by electrostatic attraction. Nanofiltration has elements of both sieving and physicochemical selectivity, the latter being on the basis of charge rejection.

Table 7.1 – Membrane materials

Material Property	Material Type	Example Materials
Chemically homogeneous	Ceramic	clay, silicate, graphite, metals
Physically isotropic	Polymeric	extruded silicone rubber
	Stretched	PTFE, polyethylene polypropylene
	Etched polymeric	polycarbonate
	Ion-exchange	functionalised polymeric materials
	Symmetric microporous	cellulose derivatives
Physically isotropic, chemically heterogeneous	Supported liquid	hydrophobic liquid in silicone rubber
Physically anisotrophic chemically homogeneous	Integral asymmetric microporous	cellulose derivatives, polyamide, polysulphone
Physically and chemically heterogeneous/ anisotropic	Asymmetric Composite	ultrathin aromatic polyamide layer on microporous polysulphone support

The subdivision of membranes into generic types based on their bulk chemical and physical properties is purely arbitrary. Microscopically, all the materials listed are, of course, heterogeneous. Membranes classified as physically anisotropic are ones whose pore size varies with membrane depth (Figure 7.2). Those classified as physically and chemically anisotropic are composite materials, comprising an ultrathin *active* or *permselective* upper layer responsible for the actual separation and the remaining part of the membrane merely acting as a support (Figure 7.3).

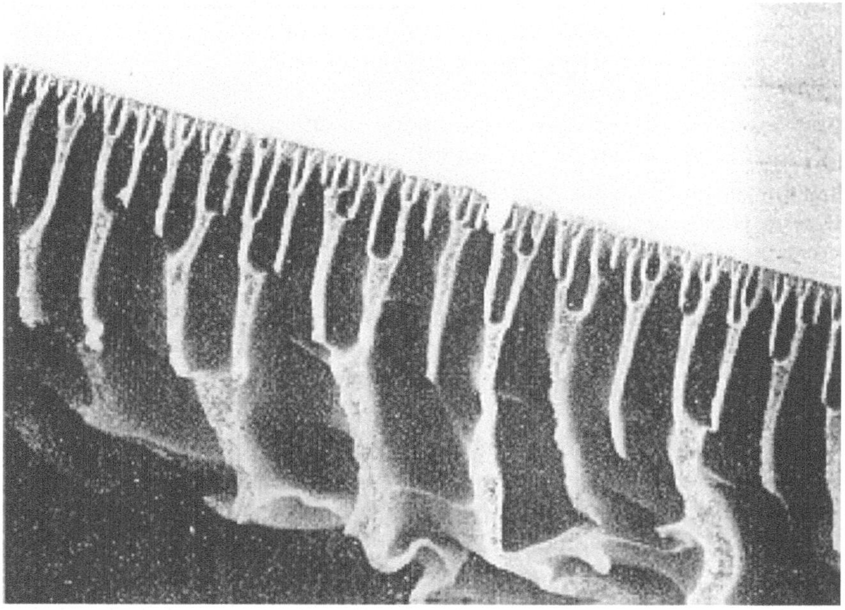

Fig. 7.2 – Anisotropic ultrafiltration membrane (courtesy of Dow Corporation, USA)

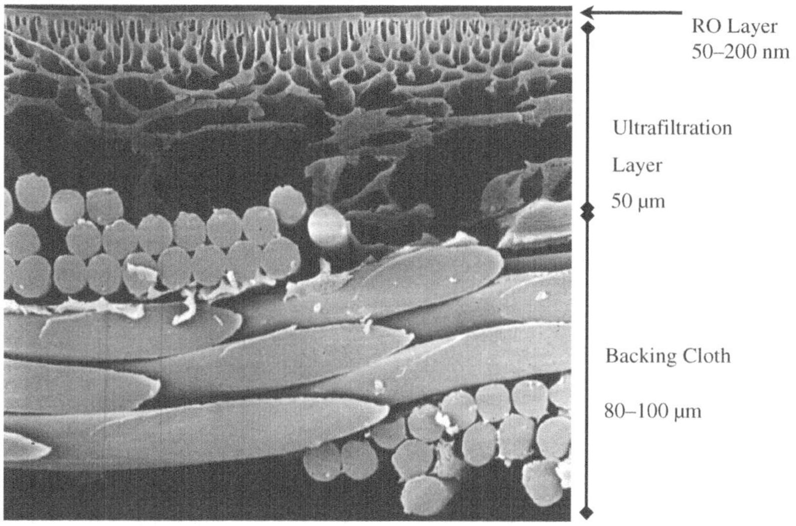

Fig. 7.3 – Composite reverse osmosis membrane (courtesy of Dow Corporation, USA)

7.4 Membrane Manufacture

Section 7.4 requires a study time of about 1/2 hour

Membrane manufacture primarily concerns the production of a porous material. The cost of the membrane is therefore dependent not only on the cost of the raw material but also on the ease with which pores of the desired size or size distribution can be introduced. Whilst some materials lend themselves well to pore production by a particular technique, others require a specific process, which may be costly.

The most inexpensive inorganic membranes are formed by pressing and sintering of fine powders to produce a porous solid of variable pore size. More isoporous ceramic or metallic membrane materials, produced by coating the sintered support with a more selective layer, are invariably more expensive than polymeric membranes of an equivalent pore size.

The pore production process for crystalline, chemically-inert polymeric materials is limited, and generally produces materials of low porosity and correspondingly high resistance to the flow. Pores of stretched or extruded membranes are introduced by stretching partially crystalline foil perpendicular to the orientation of crystallites. These membranes are inexpensive to produce but also tend to have relatively wide pore size distributions and are also isotropic (i.e. homogeneous, with the same pore size throughout the membrane). Cylindrical pores of an exact size, on the other hand, can be produced by neutron bombardment followed by etching. These production techniques can be used to produce chemically inert membrane materials for treating aggressive liquids or for sterile filtration in medical technology.

REVERSE OSMOSIS
RO membranes usually incorporate an active skin on the porous support to improve selectivity.

Symmetric microporous membranes are produced by phase inversion reaction. A solution of the polymer is precipitated in water, a process sometimes referred to as *gelation*. The technique can be extended to produce a "skin" of much greater selectivity by an additional evaporation step, producing an integral anisotropic microporous membrane. An example of a UF membrane produced by this method is shown in Figure 7.2.

The selectivity of asymmetric membranes can generally been improved by using a composite material. This allows an active skin of optimum selectivity for a particular liquid system to be chosen, whilst the porous support material and further webbing provides mechanical strength (Figure 7.3). This type of membrane is extensively used for reverse osmosis, where the active skin is usually an aromatic polyamide.

Supported liquid membranes (SLMs) have not yet been produced on a commercial scale, and currently appear to be more appropriate to gas separations where the improved mass transport of gases through liquids over that attainable in solids becomes important. Finally, ion exchange membranes can be produced either by immobilising powdered ion-exchange resins in an inert resin matrix or, more usually, by functionalisation of a homogeneous polymer film.

Section 7.5 requires a study time of about 1/2 hour

7.5 Membrane Process Parameters

In most water treatment membrane processes the membrane effectively splits the *feed* into two streams: a *permeate* or *product* which has a lower contaminant concentration, and a *retentate* or *concentrate* (sometimes also called *reject*) which contains all the contaminants removed from the permeate (Figure 7.4).

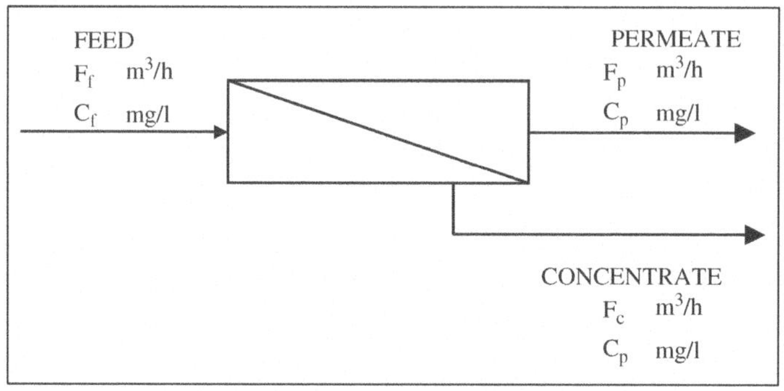

Fig. 7.4 – Generic membrane separation process

RECOVERY
The recovery of a membrane is the ratio of permeate flow to feedflow.

FLUX
Flux is flow of material (either water or slat) per unit area of membrane.

The ratio of permeate to feed flow, F_p/F_f, is called the system *recovery* or *conversion* and the removal efficiency, the ratio of the permeate to feed concentration $((C_f-C_p)/C_f)$, is called the *rejection*.

The key parameter in membrane process performance is the *flux* or flow per unit area of material passing through the membrane. In most membrane applications this is the flow of water and takes SI units of $m^3.m^{-2}.s^{-1}$ (i.e. $m.s^{-1}$), or more commonly in litres per square meter per hour (*lmh*) or $m.day^{-1}$. The flux is determined by the *net driving force* applied, the *resistance* of the membrane and the extent of membrane *fouling* and *polarisation*, as we shall see in the following sections. In most of the membrane processes used in the water industry – microfiltration, ultrafiltration, nanofiltration and reverse osmosis – the driving force is pressure difference but in the case of electrodialysis it is electro-motive force, that is voltage difference. Thus, for pressure-driven processes the energy demand depends mainly on the hydraulic resistance of the membrane and the membrane: solution interface, and can demand pressures of up to 100 bar. For electrodialysis it is dependent on the concentration of ions removed by the process.

7.6 Membrane Configuration

Section 7.6 requires a study time of about 2 hours

Membranes are normally supplied in the form of pre-assembled *modules*, which can be connected directly into pipework, or in the form of *elements* designed to fit into modules for ease of replacement. The design of the module usually aims to provide:

- a high membrane area to module bulk volume ratio;
- a high degree of turbulence to promote mass transfer;
- a low energy expenditure per unit product water volume;
- a low cost per unit membrane area and
- ease of cleaning.

No one module design can encompass all of these criteria: some of the listed characteristics are mutually exclusive. For example, promoting turbulence inevitably results in an increase in the energy expenditure. Direct mechanical cleaning of the membrane is only possible on comparatively low area:volume units where the membrane is accessible. Such module designs inevitably increase the total cost per unit membrane area. Finally, it is not possible to produce a high-membrane area to module bulk volume ratio without producing a unit having narrow feed channels, which will then adversely affect turbulence promotion.

Ex 7.2
Which of the following could not represent the units of flux?
a) litres m^{-2} hr^{-1}
b) m m^{-2} s^{-1}
c) m s^{-1}
d) gallons ft^{-2} day^{-1}
e) cm s^{-1}

There are five principal configurations that are currently employed in membrane processes are:

- pleated
- plate and frame
- spiral wound
- tubular
- hollow fibre.

7.6.1 Pleated Filter Membranes

The *pleated filter cartridge* (Figure 7.5) is used exclusively for dead end microfiltration. It is of simple and inexpensive design and construction, with the membrane being concertinaed to allow a high membrane area within a small module volume. A key property of such modules is the total dirt holding capacity, since it cannot be backflushed (i.e. physically cleaning by reversing the flow). They are thus best suited to filtration of fairly pure waters of low suspended solids levels.

7.6.2 Plate and Frame Membranes

The *plate and frame* or *stack* type is used universally for electrodialysis (Figure 7.6) and for some special applications of MF, UF and RO (Figure 7.7). These modules are generally hand-assembled, and can be disassembled if individual membranes require replacing. This makes them expensive to produce, but allow flexibility in such things as membrane selection and spacer design. Electrodialysis stacks comprise alternating cathode and anion-exchanging membranes creating alternating concentrating and desalinating compartments or "cells" which must be combined to respectively produce a concentrate and diluate stream.

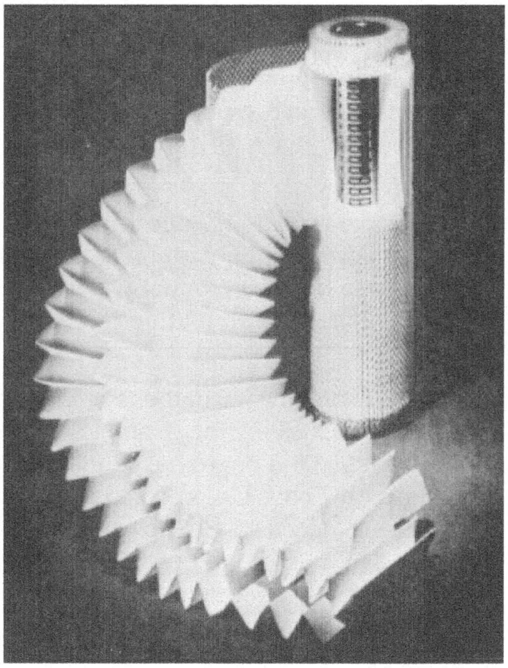

Fig. 7.5 – Pleated cartridge (courtesy of Pall, UK)

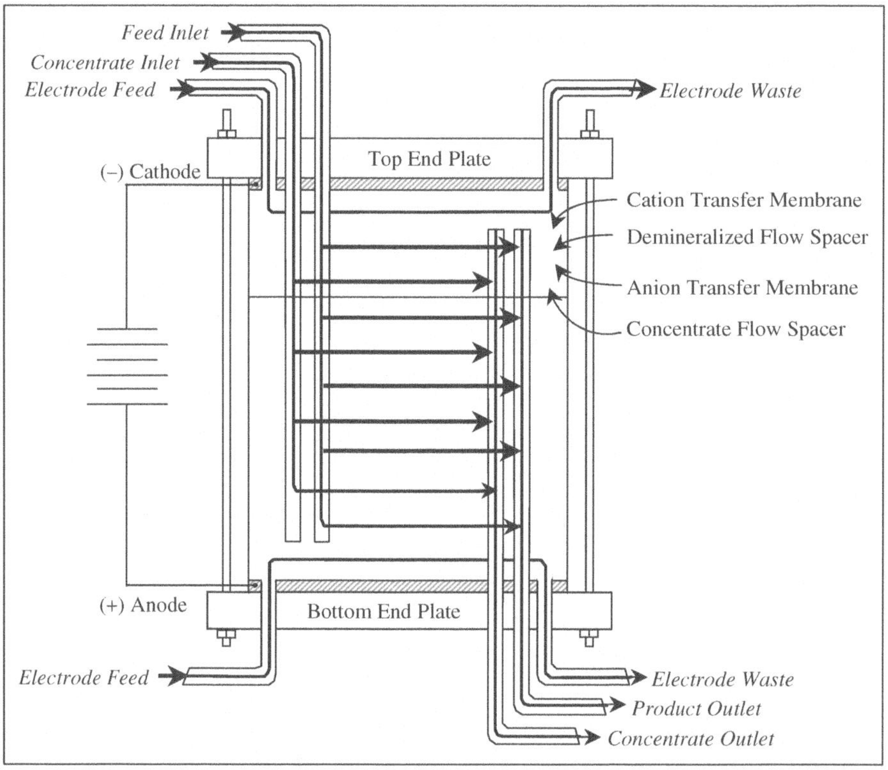

Fig. 7.6 – Electrodialysis stack (courtesy of Ionics, USA)

7.6.3 Spiral Wound Membranes

The *spiral wound* element (Figure 7.8) has become the water and wastewater treatment industry standard for reverse osmosis and nanofiltration, with many manufacturers producing standard 2½″, 4″ and 8″ diameter elements for fitting into standard pressure vessel modules. As many as 8 membrane elements can be fitted into a single pressure vessel, with the retentate outlet from one element forming the feed to the next and the permeate product from each element being combined to form a single permeate stream.

Ex 7.3
For which membrane processes is electrical current employed directly to remove contaminants, rather than just to operate a pump?

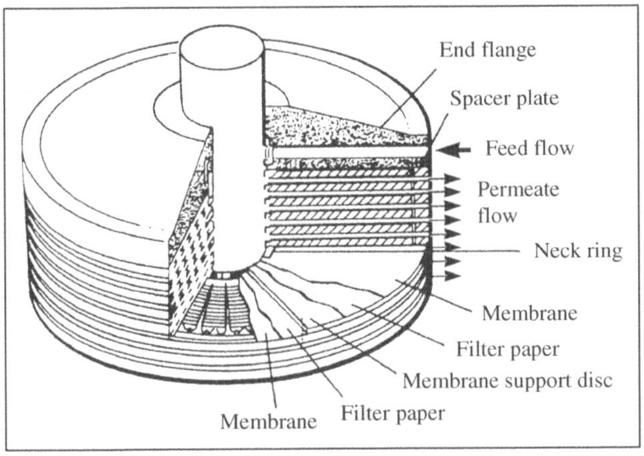

Fig. 7.7 – Plate and frame module, RO (after Ho and Sirkar, 1992)

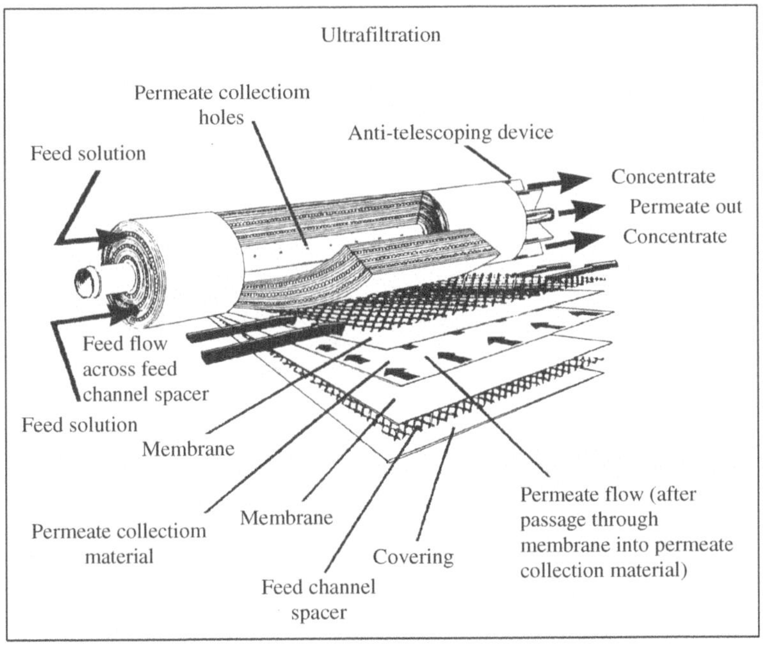

Fig. 7.8 – Spiral wound element (after Ho and Sirkar, 1992)

7.6.4 Tubular Membranes

Tubular elements, although much more expensive than spiral wound elements and with a low area to volume ratio, are often preferred for applications where the feed is particularly fouling. For example, tubular crossflow microfiltration elements are used for filtration of effluents having high suspended solids levels; tubular nanofiltration elements have been successfully demonstrated for treatment of highly coloured surface waters. Tubular membranes (Figure 7.9) have internal diameters typically in the range 5–20 mm and can be cleaned by forcing sponge balls along them to scour the internal surface.

7.6.5 Hollow Fibre Membranes

Hollow fibre ultrafiltration/microfiltration membranes of internal diameter about 0.5–1 mm are widely used for both pure and wastewaters. They have the advantage of offering reasonable mechanical integrity by virtue of the small diameter, which means that the wall thickness to bore ratio is relatively high (Figure 7.10). The finer filaments, <1 mm (as low as 40 μm for some RO elements, in which case they are referred to as *hollow fine fibre*), are usually operated out-to-in. A key advantage of HF membranes is that they can be backflushed, i.e. physically cleaned by reversing the flow. They are usually provided as tubular modules, potted so as to expose the ends of the fibre at one or both ends of the tube and packed tightly so as to provide a high membrane area:volume ratio. Recently, submerged HF membranes have been employed, where the membranes are not restrained by the walls of the module but can move freely within a tank of the water.

BACKWASHING
Membrane require periodic backwashing with either water or gas to remove trapped solids from the membrane pores.

Fig. 7.9 – Tubular membrane (courtesy of X-flow, The Netherlands)

Ex 7.4
Identify the module configurations having the characteristics listed below:

a) parallel, flat-sheet rectangular membranes; adjustable spacers.
b) high surface area to module volume ratio; backflushable.
c) multi-element module; industry standard for RO and NF.

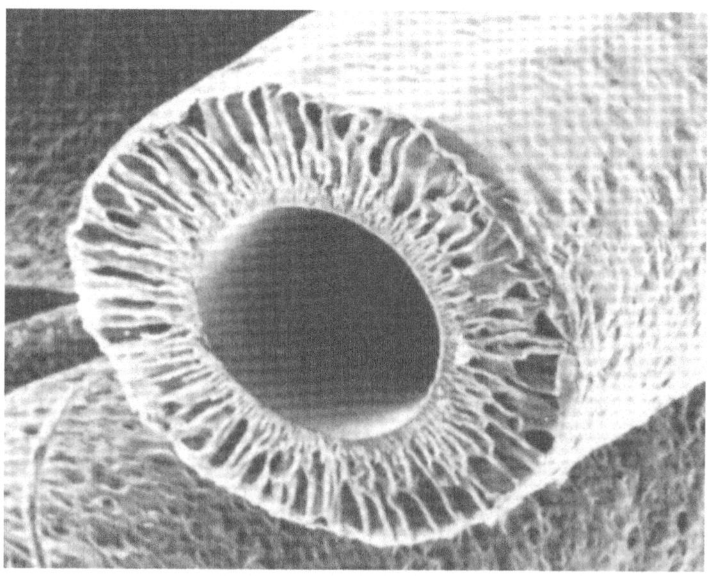

Fig. 7.10 – Hollow fibre membrane (courtesy of Memcor, Australia)

7.7 Membrane Process Operation

Section 7.7 requires a study time of about 1 hour

Membrane processes demand energy for maintaining a flow of water or ions through the membrane through the application of pressure or, in the case of electrodialysis, voltage. The passage of water through the membrane and the rejection of contaminants by the membrane inevitably leads to an accumulation of the rejected matter at the membrane:solution interface within a near-stagnant region referred to as the *boundary layer*. This accumulation is referred to as *concentration polarisation* and contributes to various phenomena which all tend to reduce the overall permeability of the membrane:solution interface (Figure 7.11). These phenomena are collectively referred to as *fouling*, and are ameliorated to a greater or lesser extent by operation at:

- lower flux (reduced conversion)
- higher crossflow velocity (and so increased turbulence), and
- higher temperature (and so reduced viscosity)

FOULING
Membrane fouling reduces the overall permeability of the membrane: solution interface.

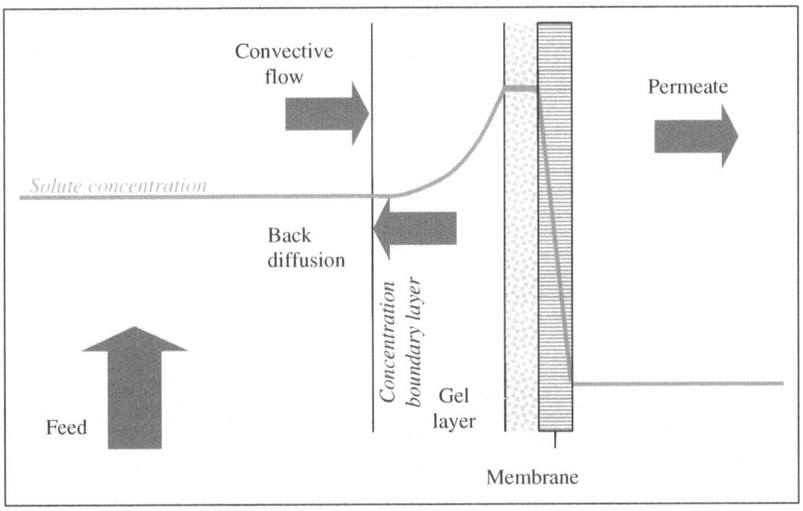

Fig. 7.11 – Concentration polarisation

Optimisation of membrane processes relies on selection of the appropriate membrane material, membrane module and operating conditions. The latter demands upon selection of a suitable flux to give a reasonably high conversion whilst still avoiding the most onerous consequences of fouling (increased downtime and reagent costs from membrane cleaning, increased pumping energy demand and, possibly, increased membrane replacement). It is nearly always the case that it is more cost effective to operate under more conservative conditions to limit fouling as much as possible. However, fouling can never be completely avoided for real water and wastewater matrices, and much developmental work is still being conducted to provide alternative low-fouling membranes and operating protocols.

7.8 Membrane Applications

7.8.1 Dead-end Microfiltration

In dead end filtration the retentate flow is effectively zero so there is no crossflow to prevent the filter surface from blocking, so dead end filtration is limited to applications where suspended solids loads are low. Membrane microfilters are widely used in the pharmaceuticals and semiconductor industries for removal of bacteria from water. Filtration to 0.2 μm removes the standard test bacterium *Pseudomonas diminuta* and water which has been filtered to this specification or better is considered sterile.

7.8.2 Crossflow Microfiltration

Crossflow microfilters can handle high levels (up to several thousand mgl^{-1}) of influent suspended solids and are now used in a wide variety of applications including:

- removal of turbidity from surface water for potable supply;
- removal of *Cryptosporidium;*
- pre-treatment of reverse osmosis feed water to reduce SDI and
- clarification of biologically treated effluent in membrane bioreactors (Figure 7.12).

MICROFILTRATION
Microfiltration is a physical separation process that can remove particles between 0.05–5 μm in size.

Fig. 7.12 – A submerged membrane bioreactor: microfiltration modules are submerged in the bioreactor tank (courtesy of Aquator Group, UK)

7.8.3 Ultrafiltration

Ultrafiltration is widely used in the pharmaceutical and biotechnology industries for selectively removing protein and similar large molecules but is gaining acceptance in many water and wastewater treatment applications including:

- colour removal in potable water treatment;
- removal of THM precursors in potable water treatment;
- removal of particulates, bacteria and viruses in ultrapure water;
- removal of pyrogens from pharmaceutical water for injection;
- removal of colloidal silica from industrial water;
- removal of COD from industrial effluents;
- clarification of biologically treated effluent in membrane bioreactors;

- removal of oil from waste water and
- recovery of pigments from paint baths.

7.8.4 Nanofiltration

Nanofiltration is not very widely used because it fits in between ultrafiltration and reverse osmosis where there are few applications. However, it has been used for:

- removal of pesticides in potable water treatment (Figure 7.13);
- softening of water by preferential removal of divalent ions and
- removal of low molecular weight COD from industrial effluents.

NANOFILTRATION
Nanofiltration is a pressure-driven process that employs both physical sieving and diffusion-controlled transport.

Fig. 7.13 – A potable water nanofiltration plant (courtesy of Viola, France)

7.8.5 Reverse Osmosis

Reverse osmosis was originally developed for desalination of brackish and sea water to produce drinking water, and this is still the largest market for the process. However, it has been readily adopted for a variety of other water and wastewater treatment processes including:

- primary deionisation of potable quality water for industrial use;
- as a two-pass process for pharmaceutical *Purified Water* and *Water for Injection;*
- polishing of semiconductor rinse water;
- removal of pesticides from potable water;
- removal of nitrate from potable water and
- landfill leachate treatment.

7.8.6 Electrodialysis

Electrodialysis was originally developed for desalination of brackish water but it its now widely used for purification of water and effluent recovery in industry. A variation called *electrodeionisation* is increasingly being used for producing high purity for pharmaceuticals manufacturing and for high pressure boiler feed.

7.9 Self Assessment Questions

SAQ 7.1 Describe the different approaches used to introduce untreated water into a membrane and how this impacts upon membrane fouling?

SAQ 7.2 Identify which of the following would not effect concentration polarisation:

a) flux
b) bulk solute concentration
c) crossflow velocity
d) solute charge
e) membrane selectivity.

SAQ 7.3 Membranes are typical supplied in the form of pre-assembled modules for ease of replacement. What other considerations should be included in the design of a module membrane?

SAQ 7.4 List the five key module configurations used in membrane processes and identify which of these have/has the following characteristics:

a) Cylindrical membrane geometry
b) Cylindrical retentate flow channels
c) Parallel retentate flow channels (i.e. flow channels bounded by parallel walls)
d) Backwashable
e) Produced solely for microfiltration applications
f) Applies to electrodialysis
g) Has the highest membrane area to module volume ratio
h) Used for large-scale desalination plant.

SAQ 7.5 List the five key membrane processes used in bulk water and wastewater treatment.

Which of these is an *extractive* process?

What type of (i) membrane, and (ii) driving force is used for this process, and what type of contaminants does it remove?

Which other membrane process, if any, removes these contaminants?

7.10 Solutions to Exercises

Ex 7.1 (c) and (e); all others are too small in molecular size to be rejected by UF.

Ex 7.2 Only (c) does not have units of volume per unit area per unit time.

Ex 7.3 Electrodialysis.

Ex 7.4 (a) plate-and-frame stack
(b) hollow fibre
(c) spiral wound.

Bibliography

Unit 7 – Membrane Processes

Cheryan M. *Ultrafiltration and Microfiltration Handbook*. Technomic Publishing, Lancaster (1998).

Ho W.S.W. and Sirkar K.K. *Membrane Handbook*. Van Nostrand Reinhold, New York, 2nd ed. (1992).

Judd S.J. and Jefferson B. *Membranes for Industrial Wastewater Recovery and Re-use*. Elsevier, Oxford (2003).

Mallevialle J., Odendaal P.E. and Wiesner M.R. *Water Treatment Membrane Processes*. McGraw-Hill, New York (1996).

Metcalf and Eddy, *Wastewater Engineering, Treatment and Reuse.* 4th ed., McGraw Hill (2003).

Nunes S.P. and Peinemann K.-V. *Membrane Technology in the Chemical Industry*. Wiley-VCH, Weinheim (2001).

Stephenson T., Brindle K., Judd S. and Jefferson B. *Membrane Bioreactors for Wastewater Treatment.* IWA Publishing, London (2000).

van der Roest H.F., Lawrence D.P. and van Bentem A.G.N. *Membrane Bioreactors for Municipal Wastewater Treatment.* IWA Publishing, London (2002).

8. Sludge Treatment and Utilisation

Contents

Aims and Objectives

This Unit covers the principles of treating sludge from water and wastewater treatment works:

After studying these notes you should be able to:

1. explain in your own words the following terms and concepts:

   thickening
   dewatering
   digestion

2. describe the different options currently available for sludge utilisation or disposal.

It is important that you are able to complete all the self assessment questions at the end of this Unit.

Essential Prerequisites

Prior to undertaking this Unit it is important that you have completed and understand the following Units:

Unit 4 Physical Processes
Unit 5 Chemical Processes
Unit 7 Biological Processes

Additional, information on the fundamentals for this Unit, refer to the following Units in *Process Science and Engineering for Water and Wastewater Treatment*:

Unit 4 Fundamentals of Microbiology
Unit 5 Fundamentals of Biochemistry
Unit 9 Introductory Mass and Heat Balances
Unit 10 Reactor Design Theory

8.1 Introduction

An end product of water and wastewater treatment processes is typically sludge; the mechanism of treating sludge depends heavily on the nature of the sludge to be treated. The presence of gross organic solids in sludge, pathogenic bacteria and toxic compounds all contaminate the sludge and so restrict the disposal options.

In this unit the different unit processes used in sludge thickening, stabilisation and dewatering in order to reduce the water content of the sludge as well as options for sludge utilisation are outlined.

Section 8.2 requires a study time of about 1/4 hour

8.2 Sludge Sources and Characteristics

Sludge can be defined as a suspension of solid materials in water. These suspensions can arise from clarification processes, primary and secondary sedimentation in sewage treatment, biological effluent treatment and various anaerobic processes.

There are three main problems associated with the handling of sludges:

- they are often composed of the substances responsible for causing the offensive character of untreated water;
- the organic constituents found in the sludge can often decompose and cause offensive odours and
- the sludge is largely water based with the *dry solids* content often only 0.25 to 12% of the sludge weight.

Inorganic sludges e.g. from mining coal, clays, minerals and fine particulates, such as chalk, sand or iron and steel dusts, are often inert and able to settle quickly forming dense deposits. These sludges are easy to treat, as water molecules are only loosely bound to the solid matrix found in inorganic sludges. However, the successful removal of an inorganic sludge from a settlement tank can be a problem due to the high density, and often digging is required. Odour problems are minimal, and only occur if the sludge becomes contaminated with organic matter and bacteria.

Ex 8.1
Why must sludge be treated?

Another type of inorganic sludge consists of metal hydroxides, formed by the addition of ferric or aluminium coagulants and sodium hydroxide or lime precipitants in water or wastewater treatment. These sludges are light and flocculant and settlement must be encouraged through tank design. In addition, the properties of this type of sludge depends heavily on the industrial or up-stream processes and there can be a wide variation in sludge characteristics. For example, iron produces a red sludge compared with the grey-brown lime-based sludge. The texture can also vary, with iron or aluminium hydrates forming a slimy or gelatinous sludge. Sludge odour is not usually a problem with industrial sludges, although sludges from algae removal clarifiers can become objectionable.

SOLIDS CONTENT
The composition of a sludge is expressed as % dry solids.

1 kg of a sludge containing 5% dry solids has 50 g solids and 950 g water.

On the other hand, odour is often a problem with sludges from biological treatment processes, and their biological breakdown must be carefully controlled to contain the release of end-products such as methane, hydrogen sulphide, nitrogen and carbon dioxide. There is a strong binding of water to the organic matter, which can make dewatering difficult. Sludge from wastewater treatment comes from various sources as shown in Figure 8.1. However, the volume of sludge produced at a typical works can be very low and often only equates to 1–2% of the total volume of sewage treated. However, the treatment of sludge often constitutes around 50% of the total works operating costs.

Primary sludge is the sludge from the primary settling tank at sewage treatment works. It is usually grey and slimy in nature and is characteristic by its offensive odour. However, this type of sludge can often be readily digested. Other organic sludges include activated (aerobic) sludges. Activated sludge will digest, but putrefaction may occur, resulting in a disagreeable odour.

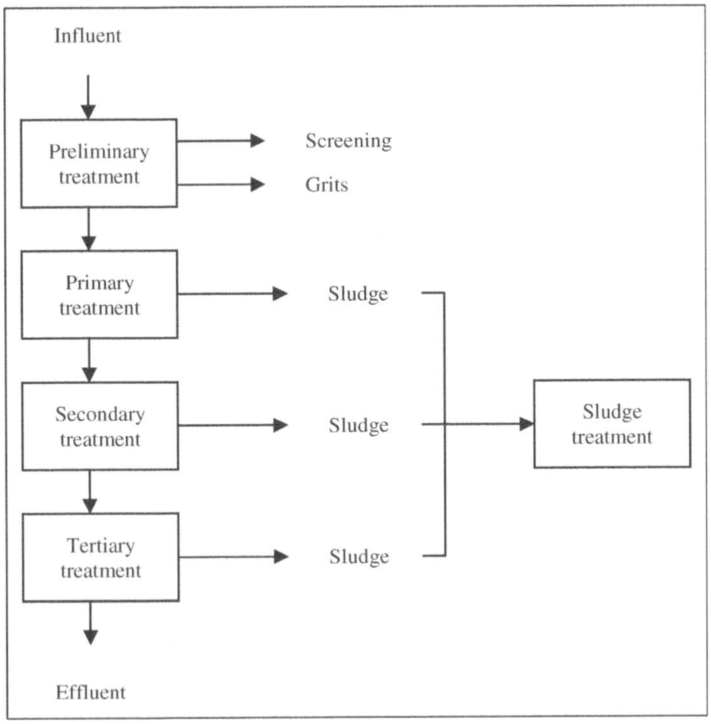

Fig. 8.1 – Sources of sludge in wastewater treatment

The quality of sludge is only partly under the control of the treatment works operator. Contaminants in the form of heavy metals and organic residues are controlled, but not always eliminated, by local trade effluent legislation. Metals and other compounds enter the sewerage system from corrosion, detergents, household cleaning products, cosmetics and pharmaceutical products, whilst the pathogen content of the waste is determined by the health of the local population.

Section 8.3 requires a study time of about 4 hours

8.3 Sludge Treatment

There are many ways sludge can be treated. Figure 8.2 shows a range of processes, which can be used in a variety of combinations.

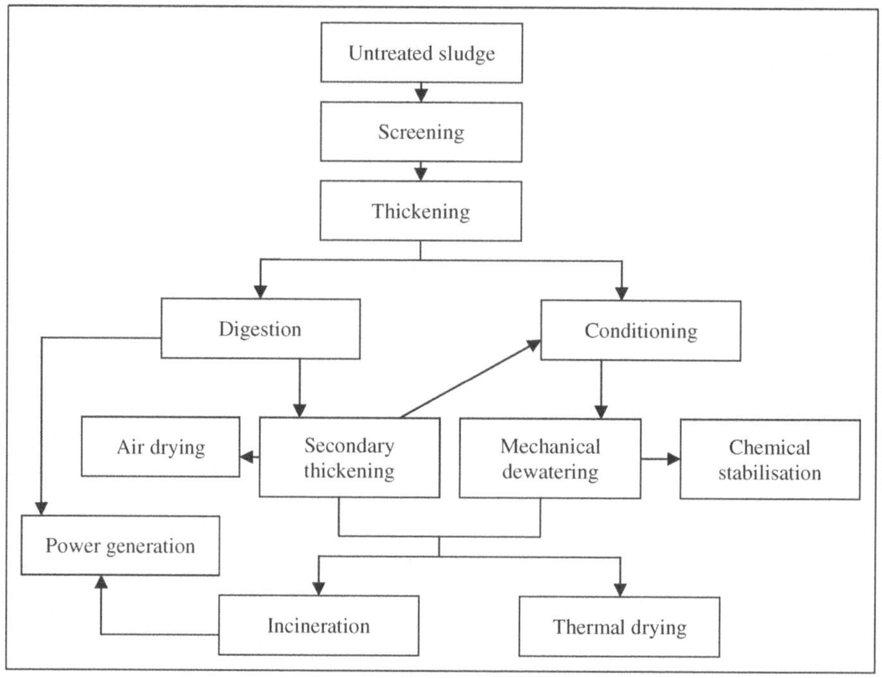

Fig. 8.2 – Outline of selected sludge treatment processes

8.3.1 Thickening

Thickening is also called *consolidation* or *concentration*. It is a procedure used to increase the solids content of sludge by removing a portion of the liquid fraction. Thickening processes are generally aimed at producing a sludge of around 4–6% dry solid but, depending on the nature of the sludge, up to 10% dry solids can be produced. For example, sludge content can change from 3% to 6% solids during the thickening process. This means an overall volume reduction of 50%. Thickened sludge is defined as 'sludge whose concentration of suspended solids has been increased by the partial removal of water but is still able to flow as a liquid' (BSI Standards, 1997).

Thickening is generally achieved by a physical processes (Unit 3), the most common being gravity, dissolved-air flotation, belt thickeners and centrifugation.

Sludge thickening is used in all areas of water and wastewater treatment:

- domestic sewage treatment;
- industrial wastewater treatment;
- potable water treatment and
- industrial water clarification.

THICKENERS
Sludge thickeners typically achieve a 50% volume reduction and aim to produce sludge between 4–6% dry solids.

Thickening is used to reduce sludge volumes either:

- to minimise handling off-site costs and
- to reduce the size of plant for further treatment.

Gravity thickening is usually achieved in circular tanks of which there are several different designs, often with the addition of polyelectrolyte to aid flocculation. *Picket fence* stirrers (Figure 8.3) are widely used for waterworks and for untreated primary and waste activated sludge from sewage works. Dilute sludge is fed to a centre-feed well and allowed to settle and compact.

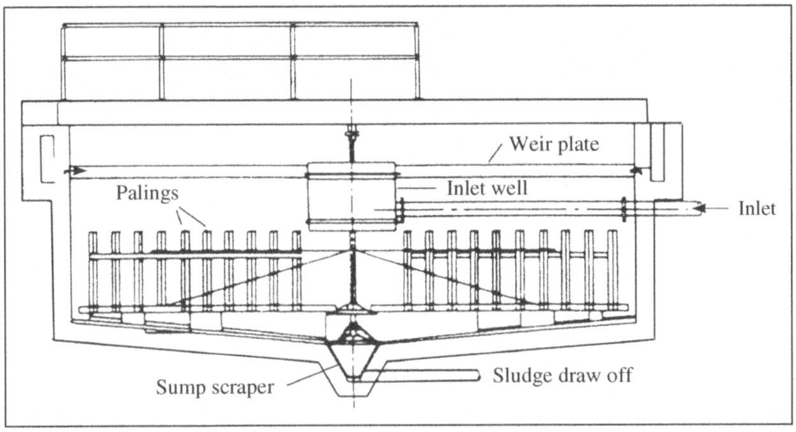

Fig. 8.3 – Picket-fence thickener (after Gray, 1999)

The thickened sludge is then withdrawn from the bottom of the tank with supernatant being returned to the works for further treatment.

There are three variations of the flotation thickening operation: dissolved air flotation, vacuum flotation and dispersed air-flotation. All three rely on the lifting of the sludge to the surface, which is then removed by a skimmer. The most common, is dissolved air flotation (see Figure 3.13). This type of system is efficient for the thickening of both untreated primary and waste activated sludges. Sludges containing metal salts from chemical treatments may also be thickened by flotation.

Ex 8.2
Why can sewage be difficult to thicken?

Gravity belt thickeners (Figure 8.4) operate by allowing the water to drain from the sludge through porous belts. Performance is flexible depending on belt speed and power and polymer consumption is low. The filtrate is good quality and the process is easily automated.

Centrifuges are also used to both thicken and dewater sludge (Section 8.3.3).

8.3.2 Stabilisation

Sewage sludge, in its raw state, is putrescible and quickly develops offensive odours. However sludge can be stabilised in order to achieve some or all of the following objectives:

- to eliminate offensive odours;
- to reduce sludge volume;
- to inhibit, reduce or preferably to prevent sludge putrefaction and
- to reduce numbers of pathogenic organisms.

SLUDGE STABILISATION
Biological stabilisation processes utilise the volatile and organic fraction of the sludge, resulting in reduced volume and increased solids content.

Sludge stabilisation achieves these objectives through acting on the organic fraction of the sludge, which serves to control the microbial population. Stabilisation methods can be biological or chemical processes. A wide range of stabilisation techniques are available. The choice is dependent on the volume of sludge to be treated and the final method of utilisation. The technologies that are available for sludge stabilisation include: lime stabilisation; heat treatment; anaerobic or aerobic digestion; and composting.

Fig. 8.4 – Belt thickener (courtesy of Anglian Water, UK)

8.3.2.1 Aerobic Digestion

Aerobic digestion is applicable to primary, co-settled sewage sludge and secondary sludges. It is the partial oxidation of sludge solids by aerobic micro-organisms under aerated conditions.

Aerobic digestion is frequently carried out in open tanks aerated by diffused air, venturi aerators or surface aerators. The capital cost of the plant is, therefore, relatively low by comparison with anaerobic digestion (Section 8.3.2.2) and is much simpler to engineer.

AEROBIC DIGESTION
Aerobic digestion of sludge typically achieves sludge volume reduction of between 30–50%.

The required retention time can be 10–15 days and the oxygen requirement is around 0.6–1.0 kg O_2 per kg dry solids. This means energy consumption for aeration and mixing is high, typically about 2 kWh per kg dry solids.

Typical sludge volume reduction is around 30–40%. Mesophilic operating temperatures are in the range 15–30°C, which is not sufficient to ensure pathogen reduction. Thermophilic aerobic digestion, however, operates in a similar manner to standard mesophilic aerobic digestion, but, operates at temperatures exceeding 70°C and therefore if operated appropriately can ensure both sludge stabilisation and pasteurisation.

8.3.2.2 Anaerobic Digestion

Anaerobic digestion is applicable to primary, co-settled and secondary sewage sludge's. It is the biological decomposition of sludge solids to produce methane and carbon dioxide.

Sludge digesters (Figure 8.5) are normally operated in the mesophilic temperature range i.e. 35°C, although some digesters are designed to operate in the thermophilic range i.e. 55–60°C. (*Process Sciences and Engineering for Water and Wastewater Treatment* – Units 4 and 5). Alternative configurations are two-stage thermophilic/mesophilic systems.

Sewage sludges are digested to stabilise and improve the product by:

- reducing the sludge volume;
- reducing the pathogen content;
- reducing odour problems;
- solubilising nitrogen and
- producing methane gas.

Fig. 8.5 – Sludge digesters (courtesy of Anglian Water, UK)

Ex 8.3
What are the main by-products of anaerobic digestion?

Anaerobic digestion consists of three different biochemical phases (Unit 6). Firstly, enzymatic hydrolysis degrades organic matter to produce amino acids and ammonia and volatile acids (VA), which are then broken down by acetogenic bacteria to form more acetic acid, hydrogen and carbon dioxide. Methanogenic bacteria then produce methane and carbon dioxide. The methanogenic stage results in mass transfer out of the digester and a lower overall solids content. When the digestion process is operating successfully there should be no significant build up of volatile acids (typical concentration in the digester should be approximately 200 mg l^{-1}) and the pH of the digesting sludge should remain around 7–7.5.

The following are basic types of anaerobic digestion:

- low rate digesters;
- single stage *high rate* digesters and
- two - stage digesters.

Digesters are usually operated in a fed-batch mode such that a fixed volume of digested sludge is withdrawn once a day or more frequently immediately followed by replacing with the same volume of raw sludge at the same temperature as the digesting sludge. The efficiency of anaerobic sludge digesters can be improved by increasing the solids content of the feed sludge. Raw or co-settled sludge can be thickened to twice the starting solids concentration in tanks prior to pumping to the digesters.

In *low* or *standard rate* systems, digestion is carried out in a single process. The sludge is pumped in near the centre of the vessel (Figure 8.6). The sludge stratifies with a scum layer at the top, supernatant below this and then the actively digesting sludge with digested sludge at the bottom to be drawn off. Limited mixing is provided by the heating of the sludge. This is often achieved by passing the sludge through an external heat exchanger. The limitation caused by the minimal mixing has resulted in the main use of this process in small installations.

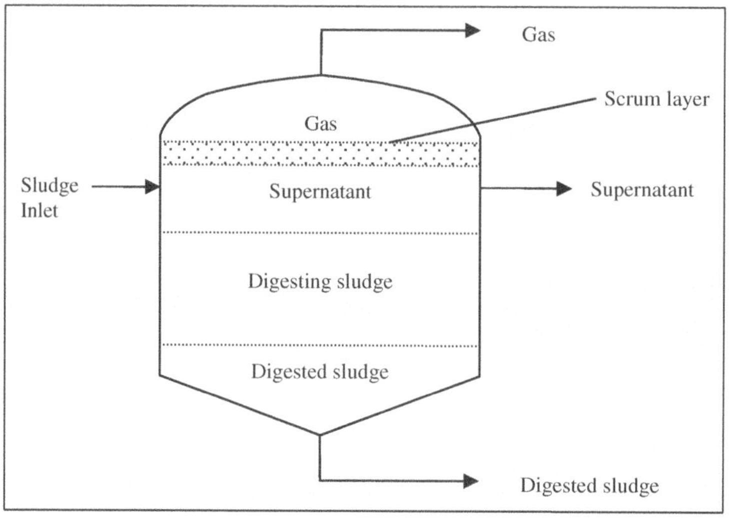

Fig. 8.6 – Low rate digester

High rate digesters are single stage, completely mixed units (Figure 8.7). These can be operated at organic loadings roughly double those found in the low rate systems. There is a decrease in the length of time required for sludge digestion, compared with the low rate digester.

Two - stage digesters are preferred to low rate digesters as they require less overall reactor volume (Figure 8.9). In the first stage the contents are mechanically mixed to ensure better contact between bacteria and substrate, i.e. as in a high rate digester. This mixing can be achieved by using mechanical stirrers or recirculation of the methane gas produced. The first stage (the primary digester) acts as a completely mixed stirred tank reactor with the second stage (secondary digester) is being operated in a similar manner to a standard rate system. In effect, this is a settlement stage, or a thickening stage, and is sometimes achieved in open top tanks.

There are many different shapes of digesters. The most frequent design is a low, vertical cylinder (Figure 8.5), alternative anaerobic tanks can be cylindrical, rectangular or egg-shaped (Figure 8.9). The rectangular shape is the most unpopular due to mixing problems and the dead zones which can form in the corners. The egg-shaped tank is a recent design and is becoming more popular. It has the major advantage that there is a reduced need for cleaning.

The actual size of the digester is determined by a number of different parameters. These include the organic loading rate, the retention time and the volume of waste produced each day.

HIGH RATE DIGESTERS
High rate digesters are single stage completely mixed stirred tank reactor.

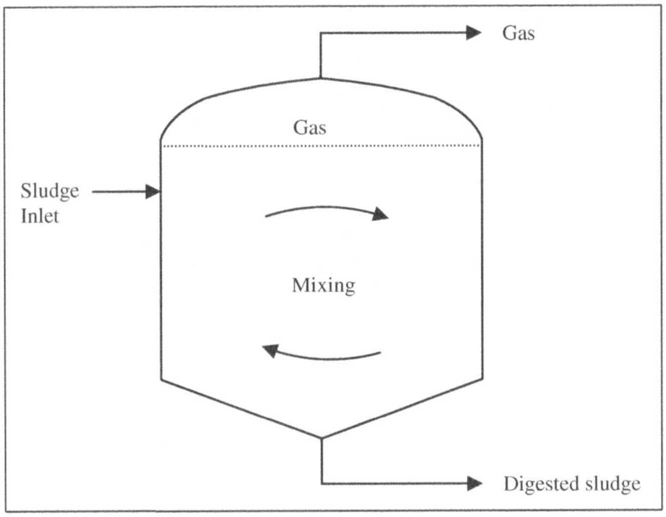

Fig. 8.7 – High rate digester

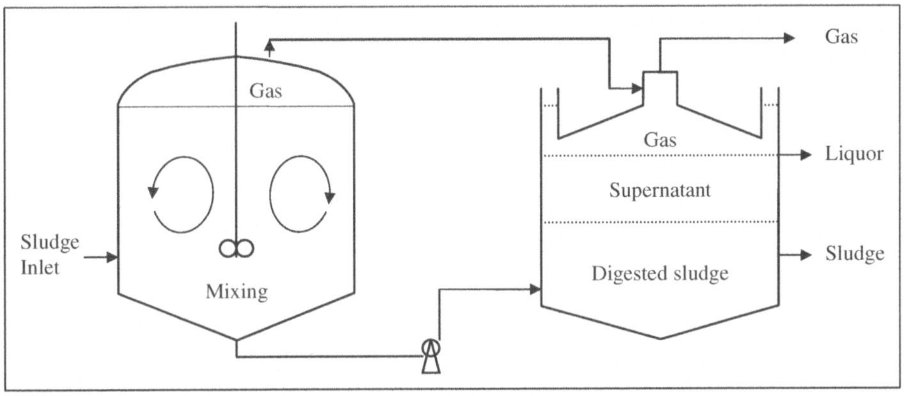

Fig. 8.8 – Two stage digester

Fig. 8.9 – Egg-shaped anaerobic digester (courtesy of Avedoere Wastewater Services, Denmark)

Mixing of anaerobic digesters is usually achieved by gas injection, mechanical stirring or pumped recirculation in order to:

- promote contact between substrate (raw sludge) and microbes;
- eliminate dead zones to reduce grit and scum;
- maintain uniform temperature and
- allow methane to be driven off.

Biogas production is dependent on temperature but is typically in the range 200–500 litres of gas per kg organic matter digested. Approximately 65–70% of the gas produced is methane and therefore has the potential to be used as an energy source. This gas is often collected under the top cover of the digester, however, a number of different gas collection systems exist. In all cases, there must be no inflow of air or escape of methane in order to prevent explosions and the inhibition of methanogenesis.

Although sludge digestion can be achieved under a range of temperatures, including psychrophilic, mesophilic and thermophilic, the most common temperature used is in the mesophilic range. This produces a good rate of digestion and methane production. In order to maintain this temperature it is necessary to provide heating for the reactor either by using the biogas or an external energy source.

Retention times are typically:

- low rate digesters 30 to 60 days;
- high rate digesters 20 to 25 days and
- two - stage digesters 12 to 16 days in the first stage followed by 3 to 4 days in the second stage.

Mesophilic anaerobic digestion (MAD) with a retention time of 12 or 15 days at 35°C does not result in a pasteurised sludge product although pathogen reduction does occur. Therefore, it is proposed for example to classify MAD treatment as producing a treated sludge product within the UK Safe Sludge Matrix or as a conventional treatment within the draft EU Working Document on Sludge (Section 8.4.2). This means that some cropping and land use restrictions are required when this sludge is applied to land.

MESOPHILIC DIGESTION
Mesophilic anaerobic digesters (MAD) are the most common sludge digester, producing good rates of digestion and methane production.

A two-stage thermophilic/mesophilic digestion system on the other hand does result in an advanced treated product if operated according to proposed guidelines (Thermophilic anaerobic digestion at a temperature >53°C for 20 hours as a batch, without admixture or withdrawal during treatment).

8.3.3 Dewatering

Dewatering is the reduction of the moisture content to obtain a solids concentration of at least 15% or more, and hence reduce the volume, of the sludge to:

- reduce transport costs;
- improve sludge handling characteristics;
- reduce odour problems and
- reduce landfill leachate production.

Dewatering is normally preceded by thickening and conditioning. Sludge conditioning is the addition of chemicals to aid flocculation and water separation. Although an expensive process due to the cost of the chemicals required it is cost effective due to the increased solids content of the sludges produced. There are numerous chemical conditioners available including lime, ferrous sulphate, ferric chloride, organic polymers and polyelectrolytes. Sludge can also be physically conditioned by heating or freezing for example. However, these methods are less popular than chemical methods.

DEWATERING
Dewatering is a physical process that reduces the liquid sludge to a solid.

Dewatering characteristics are difficult to predict. The degree of difficulty increases as the proportion of secondary sludge increases with activated sludge being more difficult than conventional humus sludge. Co-settled sludge is more difficult than new sludge and older sludges are worse than fresh sludges. Dewatered sludge is defined as that sludge 'whose water content has been significantly reduced by natural or mechanical means such that it no longer flows as a liquid'. Dewatered sludge is therefore commonly known as cake.

A number of techniques are available for the dewatering process, and all these processes must take account of:

1. the dry solids content of the cake;
2. the chemical content of the cake;
3. solids capture efficiency;
4. changes in sludge characteristics;
5. labour and maintenance costs;
6. capital costs and
7. the final utilisation or disposal route.

CONDITIONERS
Conditioners aid dewatering by improving the filtration characteristics of sludge.

Techniques available include filter and belt filter presses, vacuum filtration; drying beds and centrifugation. Both the belt filter and centrifuges are continuous processes.

In the filter press (Figure 8.10) dewatering is achieved by forcing the water from the sludge under high pressure. While a filter press can achieve high concentrations of cake solids, good filtrate clarity and a high solids capture, the mechanism is very complex, requiring high chemical and labour costs. In addition, the filter cloth has a limited life. The filter press is used as a batch process.

The press consists of several chambers formed between recessed plates with a filter cloth over the surface of each plate (Figure 8.11). The plates are held together hydraulically with sufficient force to withstand the applied filtration pressure. The conditioned sludge is pumped into chambers through feed holes in the plates and as the pressure increases, then dewatering occurs. The filtrate escapes through ports in plates while the cake is held between the plates. The plates are then separated and the cake is removed. The cake thickness can vary from 25–38 mm and the moisture content from 48–70%. The time for each cycle is normally between 2 and 5 hours, however, the cycle depends on type of sludge. The filter process is primarily used for raw sludges, but, is equally successful with digested or mixed sludges.

The belt filter press (Figure 8.12), which is also used for sludge thickening, is a continuous-feed sludge-dewatering device which generally performs well on biological sludge's but less well on inorganic sludge's.

Fig. 8.10 – Filter press (courtesy of WRc, UK)

Ex 8.4
What is the difference between thickened and dewatered sludges and what are the typical dry solids for both?

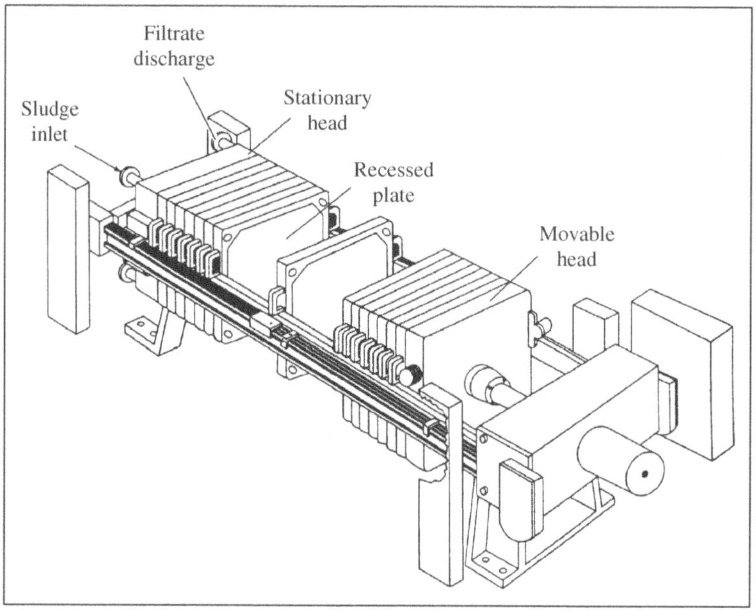

Fig. 8.11 – Recessed filter plate (after Gray, 1999)

BELT FILTER PRESS
Belt filter press is a continuous dewatering process, which squeezes excess water from the sludge between two belts.

The sludge is chemically conditioned prior to being fed onto the press. Once on the press, the majority of the free water is removed in the draining zone, by gravity. An increasing pressure is then applied in the initial compression zone, where the sludge is squeezed between porous cloth belts. The sludge is then subjected to shearing forces by a number of rollers, which release any remaining water from the sludge. The final sludge cake is removed from the belt by scraper blades. A water wash cycle is used to clean the belts.

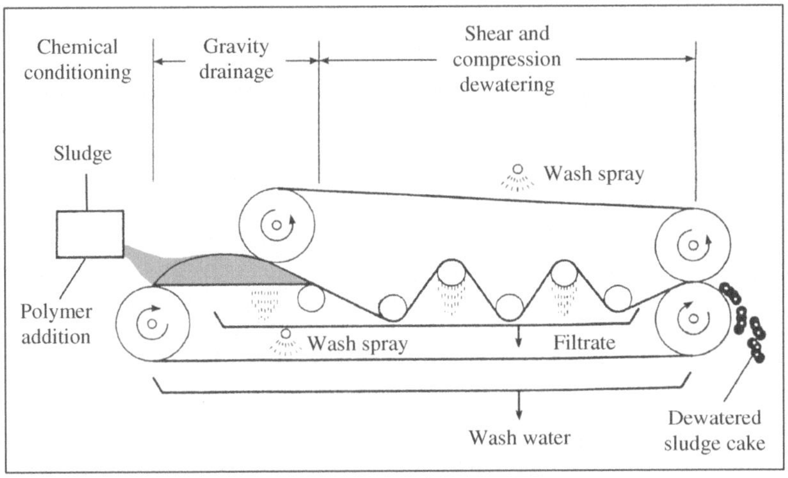

Fig. 8.12 – Belt filter press (after Metcalf and Eddy, 2003)

Filter presses produce the driest cakes, typically in the range 30–40% dry solids, but pressing is a labour intensive batch operation.

For this process, there are safety considerations to be taken into account; adequate ventilation must be provided to remove gases such as hydrogen sulphide. However, the system is compact and requires a low maintenance with a minimal operator attendance. The belt filter press is a continuous process.

Centrifugation can be applied to both thickening and dewatering and performs two functions: clarification and thickening with sludge particles being settled under centrifugal forces. The most widely used type of centrifuge in sludge dewatering is the *solid bowl scroll* type (Figure 8.13), also called a *decanter* centrifuge. Centrifugation works well on both organic and inorganic sludges.

Fig. 8.13 – Solid bowl centrifuge (courtesy of Severn Trent Water, UK)

CENTRIFUGATION
Centrifuges can be used to thicken and dewater water and wastewater sludge's to a dry solids content of 18–25%.

The centrifuge, shown schematically in Figure 8.14, consists of a horizontal elongated bowl, tapered at one end. Sludge is introduced continuously into the tapered end of the unit and the solids concentrate on the periphery.

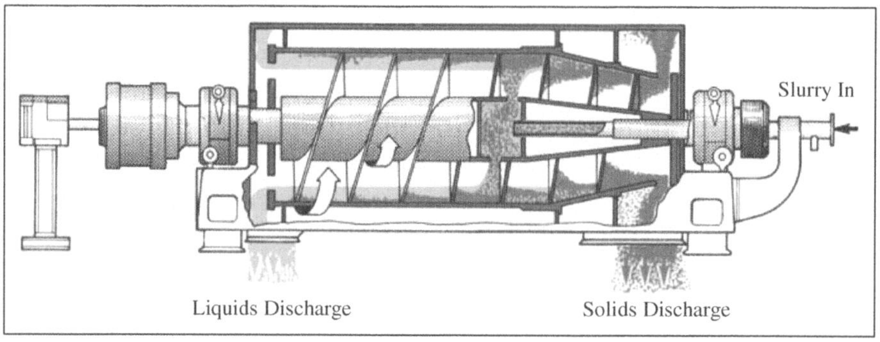

Fig. 8.14 – Sold bowl scroll centrifuge (courtesy of WRc, UK)

A helical screw spins at a slightly different speed, moving the accumulated sludge towards the tapered end, where further solids concentration occurs. The sludge is then discharged. An increase in the bowl length can increase the settling area and the retention time. This in effect will increase clarification. The pond depth can also be manipulated. A reduced pond depth followed by more beach will result in a drier cake. The bowl speed can be altered to increase the G forces, and hence improved drying of the sludge is found. By manipulating the bowl speed, a compromise can be found to achieve the optimum clarification, throughput, cake dryness and maintenance costs. Any adjustments to scroll speed will affect the centrate and cake dryness. A lower differential results in a dryer cake, while a higher differential forms a wetter cake.

The centrifuge is capable of producing cake at about 18–25% dry solids and has the advantage of being an automatic process. However, maintenance and power costs can be substantial and therefore, this process is usually found in large plants, where space is limited and skilled operators are available.

8.3.4 Thermal Drying

Thermal drying processes use heat to remove moisture from sludge by evaporation that produces a granular/pelletised product with solids content around 90% dry solid. The product can be bagged and sold as a fertiliser, although operating costs are high. This product is also pasteurised because of the high temperatures utilised.

THERNAL DRYING
Thermally dried sludge can be sold as fertiliser in either a granular or pelletised product.

There are two main types of thermal drier – *Direct or Indirect systems*. In the direct drier, hot combustion gases (from a boiler) are brought into contact with the sludge. Large quantities of exhaust gas are produced which require treatment. Typical direct drier systems include the flash drier, the rotary drum drier and the belt drier. Indirect driers use heat exchange surfaces to avoid direct contact between combustion gases and the sludge, which reduces the exhaust gas treatment requirements. Most indirect driers are closed loop systems of two generic types (single stage and dual stage). In these driers a fraction of the dried sludge product is recycled with the feed sludge so that the solids content of the sludge inside the drier is maintained above the *critical solids content* (at which hot sludge proves to be highly adhesive). This critical solid content is typically 50–60% dry solids. Indirect driers are used more commonly for sludge drying due to their energy economy, reduced gas emissions and reduced fire risk.

8.4 Sludge Utilisation

Section 8.4 requires a study time of about 1 hour

The need to manage the final utilisation or disposal of sludge as part of the overall management of a sewage works has existed since the mid-nineteenth century. However, a 'laissez faire' approach did prevail in the early days, but, now very closely regulated management practices have developed.

The constraints on sludge outlets are becoming tighter and it is essential that for each situation sustainable, secure and economically-acceptable combinations of treatment and utilisation are identified. Detailed strategic studies are now usually completed using a number of commercial available computer optimisation models (e.g. the Biosolids Manager by Salmon; WISDOM; DISPEL etc.) and a study of the Best Practical Environmental Option (BPEO). A BPEO study usually comprises a number of stages: including the collection of data on sludge production and quality, identification and evaluation of all possible treatment and utilisation options and the selection of the most appropriate options.

Sludge management must also link with the waste hierarchy in which methods to minimise and recycle are preferential to disposal and the government strategy for achieving sustainable waste management in general which is to:

- reduce the amount of waste produced;
- make the best use of the waste produced and
- minimise the risk of environmental pollution and harm to health.

Therefore, recycling sludge to agriculture is preferred to incineration, which in turn is preferred to landfill. However, landfill will continue to be a disposal outlet for sludge, particularly for ash from sludge incinerators, but due to government landfill tax initiatives (being introduced in many countries) it will become increasingly expensive and will add to the pressure to divert to utilisation options where feasible.

8.4.1 Land Utilisation

The main options for land utilisation of sludge are:

- Agriculture:
  - Fertilisers
  - Soil Improver
  - Topsoil substitute
- Reclamation/Land Restoration;
- Silviculture;
- Forestry;
- Amenity;
- Horticulture Products:
  - Compost
  - Pellets

Land utilisation allows nutrients to be recovered and offers a safe disposal of organic matter. It is a low cost, low technology option and is compatible with "green" policies. For example, in the UK, agricultural utilisation of sewage sludge represents the most important outlet.

LAND UTILISATION
Land utilisation of sludge provides a supply of nutrients as well as improving the soil structure and water holding capacity.

Sludge applied to land (Figure 8.15) acts as a soil conditioner, improving the soil structure and water holding capacity as well as providing a supply of nutrients such as P and N. However, sludge application to land can also have adverse effects, which could result in increased soil metal concentrations, nutrient leaching and possible pathogen exposure for example. Therefore, owing to the broad range of environmental protection issues that may impact on it, the use of sludge in agriculture is now a highly controlled and scientifically managed operation which is subject to a number of guidelines and regulations in different countries. These relate to the amount, timing, location and quality of sludge applied.

Fig. 8.15 – Agricultural application of wastewater sludge products (courtesy of Anglian Water, UK)

Sludge therefore must be:

- treated (conventional or advanced treatment) to reduce or eliminate pathogens. No untreated sludge applications are allowed;
- of appropriate quality in terms of metal, organic compound and dioxin concentrations and applied at rates preventing any excessive build up of these substances;
- applied at rates not exceeding crop nutrient (Nitrogen and Phosphorus) requirements e.g. 250 kg total N ha^{-1} y^{-1};
- applied in recommended locations and times using appropriate methods so as not to contaminate water courses or cause any other pollution risk;
- applied to recommended crops only and follow appropriate harvest intervals when required for conventionally treated sludge and
- analysed for composition and properties of sludge in relation to agronomic parameters and metals with all information regarding application rates and times for farms receiving sludge kept as well.

These sludge application controls vary between countries and are normally incorporated in the legislation or implemented in best practice codes and guidance.

In Europe, for example, the current position regarding sludge utilisation to land is regulated under the EC Directive (86/278/EEC) on the use of sludge in agriculture, which aims to protect soils and human and animal health by setting minimum standards for metal levels in biosolids and in soil to which it is applied. This directive is currently in the process of being updated to a concept similar to the US EPA 503 Regulation of conventionally treated and advanced treated sludge (i.e. Class A and B sludge). This classification of sludge's is with particular reference to pathogens. As is the current UK's 'Safe Sludge Matrix' agreement with the British Retail Consortium (BRC) that bans the use of untreated sludge on agricultural applications and the surface spraying of treated sludge to grazed grassland. Further requirements of the UK updated code are the incorporation of the principle of sludge treatment process control by means of Hazard Analysis and Critical Control Points (HACCP) instead of the traditional end-of-pipe command and control approaches. This aims to identify and put in place stringent process controls at critical points to prevent unacceptable risk from passing through the production process.

8.4.2 Fuel Options

Although sludge to land is heavily regulated and controlled it continues to attract adverse media and public perception. As a result of this and increasingly stringent legislation, there is a growing interest towards alternative utilisation options including thermal technologies. Some technologies currently under consideration are gasification, co-combustion in coal fired power plants and pyrolysis or 'Oil from Sludge'.

Primary products produced from these technologies include char, oil and fuel gas which can then be utilised to produce gasoline and diesel, methanol or power and heat depending on the processing technology applied.

THERMAL TECHNOLOGIES
Thermal technologies such as gasification, pyrolysis and co-combustion can be used as alternative biosolids utilisation routes in order to produce energy.

Fuel based options can include:

- Gasification;
- Pyrolysis and
- Co-combustion.

Whilst not as "green" as land utilisation, this is a continuous, *carbon dioxide neutral* option which produces energy and, if operated on site, reduces transport costs.

The conversion of sludges to special products in building materials for example will probably develop further in the future; however, it is unlikely to make a significant impact on sludge use in the near future.

8.8 Self Assessment Questions

SAQ 8.1 In wastewater sludge treatment outline the main sources of sludge and give a brief description of each sludge type.

SAQ 8.2 What sludge treatment processes would you choose for a works receiving both primary and waste activated sludge?

SAQ 8.3 List the common problems associated with sludge utilisation to land?

SAQ 8.4 Compare the environmental impacts of sludge utilisation to agricultural land to landfill and incineration sludge disposal routes.

SAQ 8.5 Describe how you might be able to improve the digestibility of surplus activated sludge (SAS) by bacteria in anaerobic digestion.

8.9 Solutions to Exercises

Ex 8.1 Reduce risks to public health; Reduce odour nuisance; Reduce sludge volume; Improve sludge stability; Preparation for final utilisation or disposal route.

Ex 8.2 Due to small size of sludge particles resulting in small spaces, which can retain water. Some of this water also exists as a 'gel', which makes its separation form the solid particles difficult.

Ex 8.3 Methane, carbon dioxide, water, ammonia, hydrogen, hydrogen sulphide.

Ex 8.4 Thickened sludge can flow (*c.* 6% dry solids). Dewatered sludge cannot (*c.* 25% dry solids).

Bibliography

Unit 8 – Sludge Treatment and Utilisation

McFarland M.J. *Biosolids Engineering*. McGraw-Hill, New York (2000).

Spinosa L. and Vesilind P.A. *Sludge into Biosolids*. IWA Publishing, London (2001).

9. Odour Management

Contents

Aims and Objectives

This Unit covers the principles of managing odour emissions from wastewater treatment works:

After studying these notes you should be able to:

1. explain the principles involved in the formation of odour emissions from wastewater treatment processes.
2. understand the different techniques used for the assessment and prediction of odours.
3. describe the different abatement processes used for the treatment of odour emissions from wastewater treatment works:

    chemical treatment
    biological treatment
    physical treatment

It is important that you are able to complete all the self assessment questions at the end of this Unit.

Essential Prerequisites

Prior to undertaking this Unit it is important that you have completed and understand the following Units:

Unit 6 Biological Processes
Unit 8 Sludge Treatment and Utilisation

Additional, information on the fundamentals for this Unit, refer to the following Units in *Process Science and Engineering for Water and Wastewater Treatment*:

Unit 4 Fundamentals of Microbiology
Unit 5 Fundamentals of Biochemistry

9.1 Introduction

Wastewater treatment has always produced unpleasant odours. However, the recent increase in the number of complaints about the release of odours from sewage and sludge treatment has forced water utilities to re-think their strategies for odour prevention and control. These growing environmental concerns about odour pollution have also led governments to consider odour policies that try to regulate odour emissions from wastewater treatment works.

The control of odours has become an important consideration in the design and the gaining of planning consent for new treatment works and solving sometimes long standing odours problems at existing works. A range of abatement technologies can be used to treat sewage odours thereby preventing the release of nuisance odour into the environment. Different techniques can also be used to assess the impact that these emissions can have on a local population. These include the measurement of odour annoyance using panellists compared to the use of surrogate measurement such as H_2S.

In this Unit the principle processes involved in the measurement, assessment and treatment of unpleasant odours from wastewater treatment works are outlined.

Section 9.2 requires a study time of about 1/2 hour

9.2 Odour

Odour is described as the property of a substance that activates the sense of smell or a sensation perceived by the receptor nerves in the human nose. Odour is the most offensive form of air pollution and can be results from the action of different chemicals present in the air. It is not possible to predict the odour of a chemical even if the chemical formula is known. The sensation of odour is concentration dependent and a mixture of two or more odorous substances may give an odour quality that is dissimilar to the two components when smelled alone.

9.2.1 Odour Formation

HYDROGEN SULPHIDE
H_2S is a highly toxic gas, being fatal at concentrations exceeding 600 ppm. It has an occupational exposure limit (OEL) of 10 ppm at which concentration it is extremely odorous.

Odours associated with wastewater treatment are made up of a number of compounds and results from microbial respiration (oxidation of organic matter) and/or metabolism (fermentation of organic matter) producing a variety of sulphur and non-sulphur (nitrogens, fatty acids, alcohols and ketones) based compounds. Hydrogen sulphide (H_2S) is the most important of the sulphur based compounds formed under anaerobic conditions and when even it is not the dominant compound it is often associated with the odour. It is also a toxic and corrosive gas. Other odours, which are present in sewage, can be derived from industrial discharges to the sewer such as solvents, petrol derivatives and other volatile organic compounds (VOCs). These intermittent discharges can lead to odour problems when their react with H_2S to produce sometimes even more unpleasant odours.

9.2.2 Sources of Odours

The principle sources of odour during wastewater treatment are pumping stations, inlet works and sludge holding tanks, which are not generally covered. Screening and sedimentation are two other facilities where odour can be detected. Odour can also be released from biological treatment process such as activated sludge and tricking filters mainly due to inadequate ventilation. Sludge treatment also produces odours but is dependent on the method of treatment. The source and cause of odours in wastewater treatment are summarised in Table 9.1.

Table 9.1 – Principle sources of odours from wastewater treatment works

Principle source	Cause of odour problems
Inlet works	Receiving sewage contains intermittent industrial effluents. Operational conditions: cascades, turbulence and rapid mixing
Primary sedimentation tanks	Sludge accumulation (anaerobic conditions); long retention times; large open surface areas
Activated sludge	Inadequate aeration; stripping odours from supernatant liquors; inadequate sludge production
Biological filters	Overloading/inadequate ventilation; striping odours from supernatant liquors; inadequate sludge production
Clarifiers	Inadequate de-sludging; open topped units
Sludge treatment	Excessive turbulence; open topped units; inefficient combustion of digester gas

9.3 Odour Measurement

Section 9.3 requires a study time of about 1 hour

In order to investigate odour problems at wastewater and sludge treatment works and determine if odour abatement technologies are working correctly, odour measurements must first be quantified. Currently, the most common methods used to measure odours rely either on the use of odour assessors to determine human thresholds of detection or the use of surrogates such as H_2S for determining odour strengths. However, other methodologies can be used to assess odours such as measuring individual odorous chemicals by gas chromatography systems such as GC-MS or by directly measuring VOCs by flame ionisation detectors (FID). More recently electronic nose technology has also been used to measure sewage odours.

9.3.1 Olfactometry

Olfactometry is used to quantify the strength or concentration of odours using a panel of human sniffers. The strength or concentration of an odour is defined as the number of dilution at which 50% of the panellists can detect no odour or fail to distinguish it from odour-free air and is more commonly termed the Odour Concentration (OC), also referred to as the Threshold Odour Number (TON). Although the concentration is dimensionless the odour threshold is usually expressed as odour units per cubic metre of air (ou/m^3). The procedure used for determining odour concentration involves odour samples being collected usually in inert flexible bags and being transport to an odour laboratory for analysis. The odour samples are then measured using a dynamic dilution olfactometer, whereby a range of diluted samples are presented to a panel of 6 or more odour assessors (Figure 9.1). Each olfactometer has two sniffing ports with odour-less air being presented to the panellists through one port and diluted odorous air through the other. A range of at least five dilution steps (each differing by a factor of 2) are presented to the panellists in ascending concentrations, each panellist has to indicate which port contains the odorous air.

ODOUR CONCENTRATION
The number of dilutions at which 50% of the panellists can detect an odorous sample when compared to odour free air.

Fig. 9.1 – Olfactometry panel (courtesy of Silsoe Research Institute, UK)

CEN ODOUR STANDARD
CEN standard defines a method for the objective determination of the odour concentration of a gaseous sample using dynamic olfactometry.

Although olfactometry relates most closely to what is been detected by odour complainants, it has a number of limitations. The measurements are time consuming, labour intensive, expensive and are subject to large variations between panellists. One of the main reasons for the inconsistency in odour measurements is the lack of standards and inadequate information between different tests. Until recently no clear standardisation was available for threshold olfactometry however, a new European standard using dynamic olfactometry has been drafted by the Committee European Normalisation (CEN) working group CEN/TC254/WG2 "Odour". The CEN Standard (EN 13725) implies that repeatability (within-laboratory) and reproducibility (between-laboratory) of odour concentrations measurements under reproducible conditions should not differ by more than a factor of 3 or 4 in 95% of cases. However, as odour laboratories are often remote from the wastewater treatment work and with the increasing need to assess odours on site and in real-time, it is become necessary to be able to monitor odours continuously or be able to do spot measurements using portable equipment.

Ex 9.1
What are the principle sources of H_2S at a wastewater treatment works?

9.3.2 H_2S Analysis

H_2S is the most common odorous component associated with sewage and is often used by wastewater treatment operators as a marker for determining odour strengths. It is regarded to be easier, more reproducible and cheaper than olfactometry, when large numbers of odour concentrations are required for indicating sources of odours in terms of emission rates and estimating the impact of a wastewater treatment works on their surroundings, through the use of H_2S maps. H_2S has also been reported to be more sensitive than odour concentration measurements for measuring threshold odour concentrations and it can be monitored continuously or with spot measurements using portable equipment.

Ex 9.2
What are the advantages of H_2S measurements compared to other odour monitoring methods?

Fig. 9.2 – Jerome 631-X H_2S analyser (courtesy of Arizona Instruments, USA)

A range of field instruments are available for the detection and/or measurement of H_2S. These include either relatively inexpensive techniques such as electrochemical detectors or chemical indicator tubes or H_2S analysers (based on gold-film resistance detectors) such as the Jerome 631-X H_2S analyser (Arizona Instruments, USA) that can detect H_2S concentrations down to ppb concentrations (Figure 9.2). Since H_2S is not the only compound present in wastewater, it is sometimes not the most representative marker for indicating odour strengths. Studies comparing H_2S concentrations and odour concentrations have shown that clear relationships are not always present. Additionally, H_2S would be an unsuitable marker for measuring odours from secondary treatment, sludge drying or when odour is caused by a specific industrial discharge.

Section 9.4 requires a study time of about 1 hour

9.4 Odour Assessment

The impact that an odour can has on a local population involves the assessment and prediction the release of nuisance odours from wastewater treatment works in order to provide effective control. Different techniques can be used to predict the emission of specific compounds and/or odours from different wastewater sources. The techniques used for odour assessment include the use of H_2S maps, dispersion and odour models.

9.4.1 H_2S Mapping

H_2S mapping is a technique of taking a large number of H_2S measurements within and around a wastewater treatment works and using a surface contouring technique to produce a contour map of H_2S concentrations. Figure 9.3 shows an example of the use of a H_2S map to quantify a source.

The production of H_2S maps is usually prepared using a portable H_2S analyser. The map provides a simply visual interpretation of an odour problem associated with a wastewater treatment works, as well as a means of quantifying the impact of an odour and the diagnosis of its cause. The weakness of H_2S maps is that it is not responding to all odours that typically arise from wastewater treatment. In particular, a H_2S map will give little impression of odours from secondary treatment.

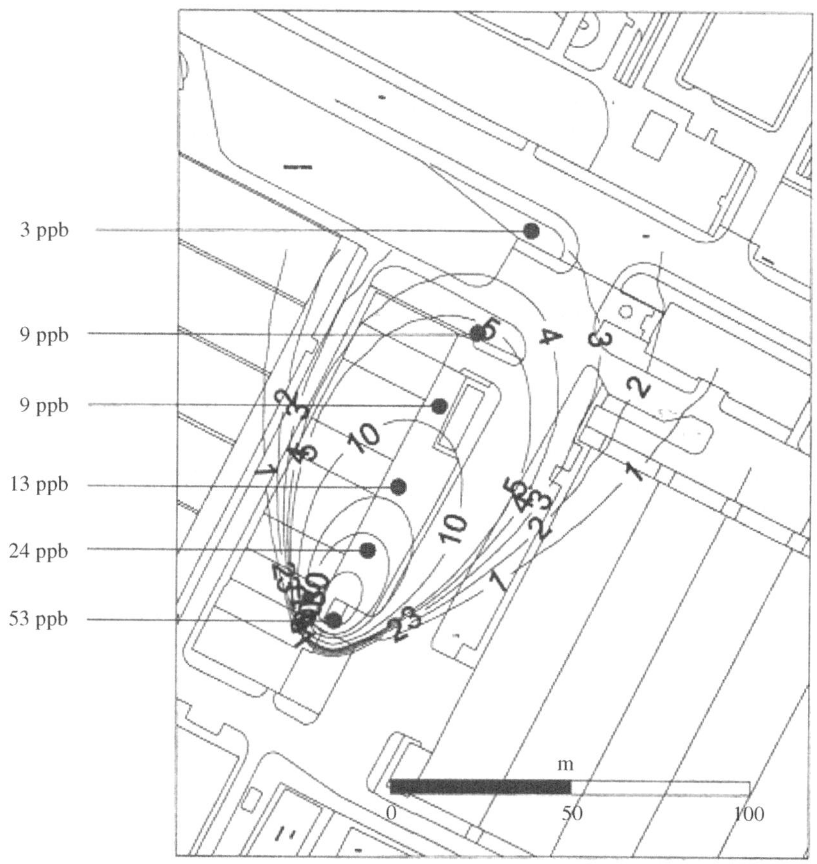

Fig. 9.3 – Use of a H_2S map for quantifying an odour source (courtesy of WRc, UK)

9.4.2 Dispersion Modelling

In order to determine the impact that odorants have on a local population, the dispersion of gaseous odorants from the source to the receptor must be determined. Commercial models are available that can predict the concentrations downwind from an odour source, these models incorporate the effects on wind speed and direction and atmospheric stability.

The most commonly applied dispersion models (such as AERMOD, ADMS and ISC) are based on Gaussian dispersion equations. These packages enable practical modelling of multiple odour sources and receptors, are designed to be user friendly and employ a graphical user interface for both model input and output. The more complex models will also allow for both point and area sources and will consider factors such as variable topography and turbulent effects caused by building and site. The output is usually in the form of the highest concentrations predicted for each receptor or some from of percentile of that predicted for the range of conditions contained in the meteorological input files.

DISPERSION MODELS
Dispersion models calculates the dispersion of gaseous odorants downwind from a source to a receptor in order to determine its impact on a local population.

Figure 9.4 shows an example of a typical odour contour plot, the model output consists of an odour contour plot, overlaid on a plan of the site and/or the surrounding area.

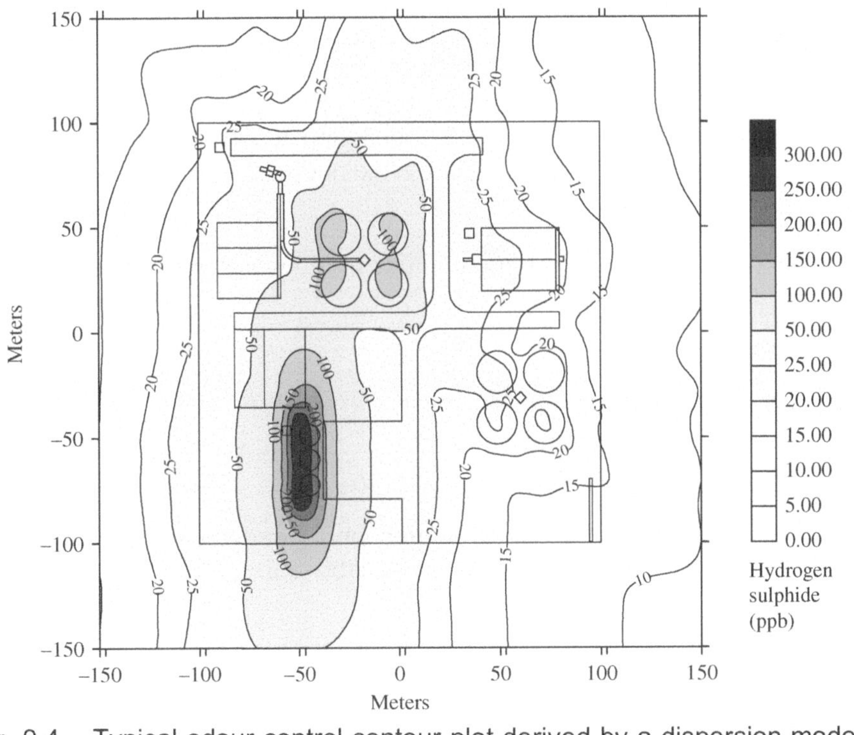

Fig. 9.4 – Typical odour control contour plot derived by a dispersion model

Section 9.5 requires a study time of about 2 hour

9.5 Odour Treatment

Once it has been determined that there is an odour problem at a wastewater treatment works and that the only way to prevent or control the impact of these odours have on there surrounding areas is by odour abatement. A decision must be made on the type of abatement systems that is most appropriate to the particular application.

Factors that will affect the choice of processes include:

- The maximum allowable odour intensity at the boundary of the sewage works;
- The flow-rate of air to be treated;
- The type and concentration of odorous compounds including variability (infrequent emissions of high odour strength potentially exceed the capacity of a control systems) and
- The space availability and zoning requirements of the proposed location.

As well as having to decide on the type of treatment system to be used on a particular odour problem. Design considerations also have to be made on the footprint or layout of the abatement system. Odours can either be drawn to a centralised abatement systems or be treated by individual systems. Sometimes, two or three stage of odour treatment are required or where there are a number of sources, pre-treatment of more highly odorous air might be required to provide greater reliability of operation and the capability to deal with a variation in loads.

Design of an abatement system must also consider operation and maintenance procedure and regular monitoring to ensure that the performance of the abatement systems is maintained and that the type of odour problem has not changed.

The main types of odour abatement in use in the water industry are:

- biofilter;
- biotrickling filters;
- bioscrubbers;
- wet ozone scrubbing;
- chemical oxidation/scrubbing and
- adsorption.

A summary of some of the advantages and disadvantages of the most common treatment systems are shown in Table 9.2.

9.5.1 Biological Odour Treatment

Biological treatment systems utilise the ability of micro-organisms develop a biofilm on the support medium to oxidise the malodorous compounds to less odorous compounds. Odorous compounds are first dissolved in moisture or liquid surrounding the media and are then oxidised directly or indirectly by the bacteria. As with all other types of biological systems, micro-organisms can be inhibited if the concentration of the compounds becomes too high resulting in substrate inhibition.

Table 9.2 – Summary of odour treatment processes

Process	Advantages	Disadvantages
Biofilters	Simple to operate Reliable Simple technology	Large footprint Difficult at very high H_2S concentrations Media degrades
Biotrickling Filters	Simple to operate Inert carrier material	Requires nutrient feed Difficult at very high H_2S concentrations
Bioscrubbers	Low operating costs Survives breakdown Good for high concentrations	Requires nutrient feed Requires process control H_2S may hamper removal of other compounds
Chemical Scrubbers	Easy to monitor and control Good removal if multi-stage Quick to re-establish after breakdown	High operating costs Storage of dangerous chemicals
Adsorption	Good for polishing Cheap to install	High operating costs Not good for high H_2S Effected by humidity
Wet Ozone Scrubbers	Meet tight standards Single stage process	Expensive May need chemical correction

The most common biological odour treatment systems are biofilters (Figure 9.5), biotrickling filters (Figure 9.6) and bioscrubbers (Figure 9.7). The principle difference being that for bioscrubbers odorants are absorbed by a circulated liquid in an absorption tower and then treated in a separate bioreactor unit, in biotrickling filters a liquid medium (such as final effluent) is circulation over the packed bed and the odorants are taken up and degraded by biofilm microorganisms on the carrier material, whereas in biofilters the gas is humidified first and then forced through the organic carrier material which is keep humidify.

BIOFILTERS
Compost, peat, bark or a mixture of these are common organic carrier material in biofilters.

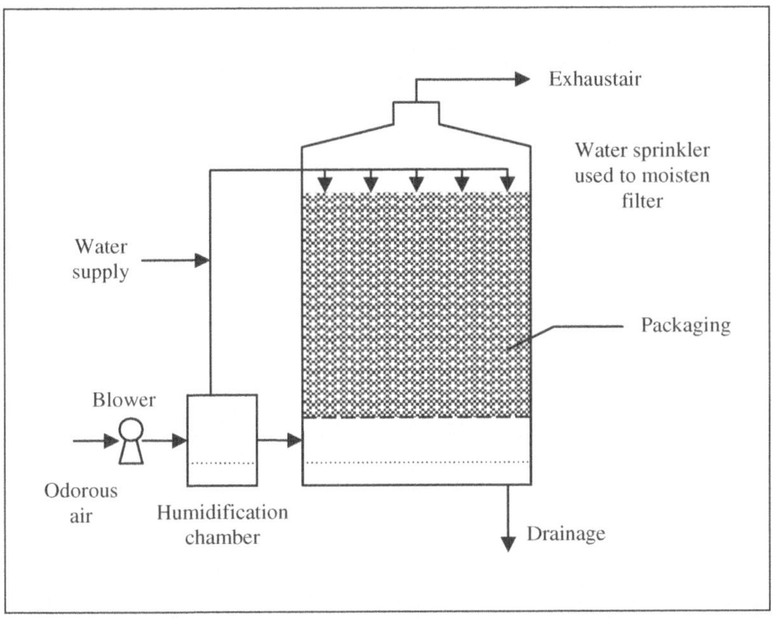

Fig. 9.5 – Biofilter

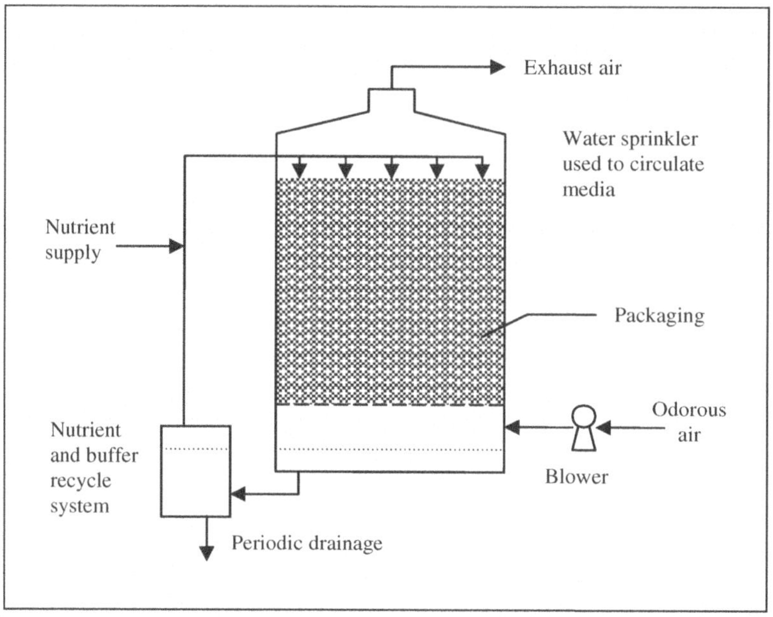

Fig. 9.6 – Biotrickling filter

BIOSCUBBERS AND BIOTRICKLING FILTERS
Wastewater effluent is a common nutrient feed used in biological odour treatment systems.

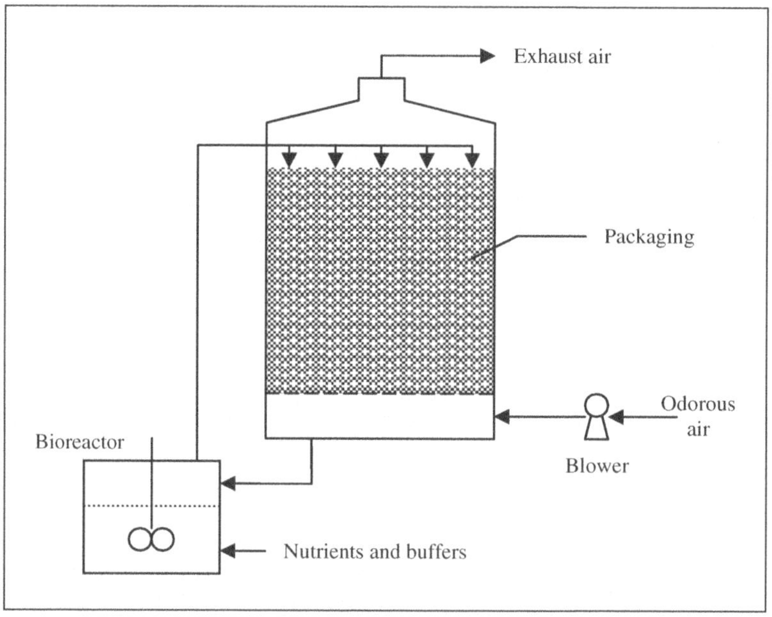

Fig. 9.7 – Bioscrubber

Some factors that affect the performance of biological systems include the correct physico-chemical environment for the growth of microbial population such as:

- oxygen availability;
- gas transfer of odorous compounds to liquid phase;
- moisture content;
- temperature;
- pH;
- macro-nutrients (carbon, nitrogen availability) and
- micro-nutrients.

If one or more of these factors are improperly applied, the performance of certain portion of the biological population can become affected and therefore limit the overall performance of the treatment system.

The performance of biological odour treatment systems is generally very good in wastewater treatment application due to their ability to treat a variety of odorous compounds. A major limitation is its inability to meet extremely high performance standards under variable loads. However, since biological processes are usually more cost-effective to operate, if is often the case that two-stage treatment system are employed, rather than using a single stage process that has a higher operating cost.

Other disadvantages to biological systems is that there cannot sometimes degrade synthetic compounds that are frequently found in industrial situations and that they occupy a relatively large footprint for the volume of air that they treat compare to chemical odour treatment.

9.5.2 Chemical Odour Treatment

Chemical scrubbing systems rely on the contact of a scrubbing chemical with the odorous gas stream. Packed tower is the most common configuration (Figure 9.8). Compounds are absorbed into the liquid medium that is circulated through the packed tower scrubber. Sorption is the primary removal mechanism; however oxidation in the liquid solution can greatly enhance the adsorption and therefore enhance removal. The standard operating configuration for a packed tower is a vertical shell with gas flowing up through packing and the liquid scrubbing solution going down through the packing system. The gas and liquid pass over the packing material to promote a large interfacial area. Liquid solutions are circulated from a collection sump and usually contain a chemical or side stream treatment option such as advanced oxidation processes (AOP) to breakdown the accumulated contaminants.

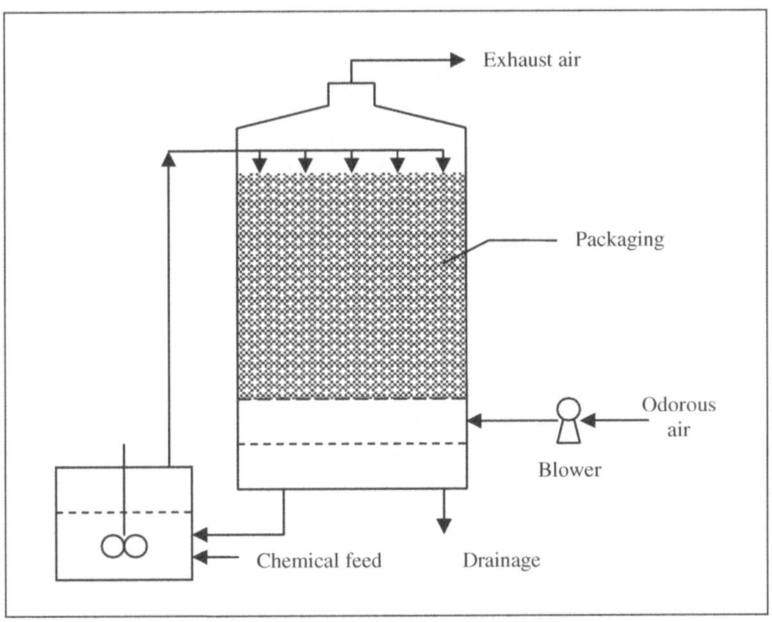

Fig. 9.8 – Packed tower with streamside chemical treatment

CHEMICAL SCRUBBERS
Chemical scrubbers often use AOP processes in breakdown contaminants in the circulation liquids.

Packed towers do an excellent job of removing H_2S, but can have problems with some organic compounds that a not water soluble. An alternative to the packed towers are mist chamber systems (Figure 9.9) that use high performance atomising nozzles that increases the interfacial area, but have less liquid solution to remove adsorbed pollutants.

Fig. 9.9 – Mist scrubber system (courtesy of Environmental Management Consulting, USA)

9.5.3 Physical Odour Treatment

ADSORPTION SYSTEMS
Activated carbons impregnated with caustics (NaOH or KOH) are the most often used adsorbents for H_2S treatment systems.

The most comment adsorption system for odour treatment is carbon based systems that consist of static beds of granular materials in vertical cylindrical columns. The absorbent accumulates a concentration of molecules from the gaseous environment. The absorbents maybe previously impregnated with a reagent that is selective for the removal or destruction of specific odorants or with a catalyst that speeds up the desired reaction.

9.6 Self Assessment Questions

SAQ 9.1 What are the two most common methods used to measure odour emissions from wastewater treatment works?

Describe how these measurements are used to determine the impact on a local population.

SAQ 9.2 What mechanisms are used to biological oxidise odorous compounds?

How does the physico-chemical environment influence the process and what factors affect its performance?

9.7 Solutions to Exercises

Ex 9.1 Primary sedimentation tanks, sludge storage and thickening and sludge digestion.

Ex 9.2 Cheaper when large numbers of samples are required compared to olfactory analysis. Potable instrumentation allows field and on-line analysis of abatement systems.

Ex 9.3 Inorganic elements such as N, P and other trace elements such as K, Ca and Mg are vital to synthesis of microbial cells.

Bibliography

Unit 9 – Odour Management

Gostelow P., Longhurst P.J., Parsons S.A. and Stuetz R.M. *Sampling for Measurement of Odours*. IWA Scientific and Technical Report No. 17, IWA Publishing, London (2003).

Stuetz R.M. and Frechen F.-B. *Odours in Wastewater Treatment: Measurement, Modelling and Control*. IWA Publishing, London (2001).

Vincent A. and Hobson J. *Odour Control*. CIWEM Monographs on Best Practice No. 2. Terence Dalton Publishing, London (1998).

WEF. *Control of Odors and Emissions from Wastewater Treatment Plants – MOP 25*. Water Environment Federation (2004).

Solutions to Self Assessment Questions

Unit 1 – Water Quality

SAQ 1.1 Examples of pathogenic organisms included bacteria, protozoans and viruses. Waterborne disease is where a pathogen is transmitted by the faecal-oral route via the ingestion of contaminated water (or other materials) by humans.

SAQ 1.2 Indicator organisms are used for the indication of presence or absence of pathogenic organisms. This results from the difficult to isolate and identify pathogens directly due to low numbers and isolation techniques employed.

Criteria – organisms must always be present when pathogens present and absent when pathogen absent; should originate from digestive tract; easy and rapid to identify; should survive longer in the environment, should be present in high numbers and not pathogenic.

SAQ 1.3 BOD – Biochemical oxygen demand: is the amount of oxygen absorbed by a sample of liquid (i.e. wastewater), saturated with dissolved oxygen (approx. 9 mg/l) and held at 20°C for the period of 5 days (BOD_5) or more. The BOD is a measure of the levels of biodegradable organics using suspended microorganisms.

SS – Suspended solids: is the concentration of insoluble material in a liquid (i.e. wastewater). Such solids are usually measured by filtration of a known volume of liquid through a pre-weighed filter paper. SS can also be inferred from the turbidity, although this method is imprecise.

Environmental discharge consent for example in the UK are based on a BOD and SS – 20 mg/l BOD and 30 mg/l SS, respectively.

SAQ 1.4 Compliance monitoring involves regular monitoring for a particular parameter in order to ensure that the water or wastewater quality complies with a particular standard or discharge consent. It is mainly achieved through sample collection and retrospective analysis in the laboratory; whereas performance monitoring enables a treatment operator to determine the effectiveness or efficiency of a process, in most case using indicator measurements that are not consent parameters.

Solutions to Self Assessment Questions

Unit 3 – Physical Processes

SAQ 3.1 The principal features of a lamella clarifier are: plates sloping at 60°, 50 mm pitch, water flow is upwards, sludge collects on plates flows down, high settlement area, short settlement path.

The advantages of lamella clarifiers are they small footprint and easy to retrofit.

SAQ 3.2 The three mechanisms of depth filtration are: direct interception, inertial deposition, diffusion, sedimentation and hydrodynamic.

Deep bed filters are cleaned by backwash with water to fluidise sand and air scouring to stir.

The advantages of a continuous sand filters are that backwash is a small continuous flow and filtration is continuous (i.e. no standby). However the disadvantages are high capital cost and complex design.

SAQ 3.3 Mechanisms of removal:
Dissolved air flotation: attachment of bubbles that change density.
Hydrocyclone: separation by difference in density of particles.
Sludge blanket clarifier: flocs are removed by being capture by other flocs which maximises the settling rates of the particles.
Wire mesh screen: particles are physically capture in the mesh.

SAQ 3.4 Lamella plate clarifier features include: plates sloping at 60°, 50 mm pitch, water flow is upwards, sludge collects on plates flows down, high settlement area and short settlement path.

Principal advantages: small footprint and ease of retrofitting.

SAQ 3.5 Three depth filtration mechanisms include: direct interception, inertial deposition, diffusion, sedimentation, hydrodynamic.

Rapid gravity filters are cleaned by backwashing with water to fluidise and air scouring to stir the sand.

Advantages of continuous filters are: backwash is a small continuous flow and filtration is continuous with no standby phase or additional filter units. Disadvantages include: high capital cost and complex design.

Solutions to Self Assessment Questions

Unit 4 – Chemical Processes

SAQ 4.1 Collision frequency and efficiency, shear rate (G value), particle size and inter-particle forces are the main factors that effect the formation of flocs in water treatment.

SAQ 4.2 Metal ions that that are used in the coagulation/flocculation influence the charge neutralisation whereas the addition of polymers effect floc binding.

SAQ 4.3 The coagulation/flocculation process removes colour, taste, odour, organic carbon and particles (turbidity).

Aluminium and iron salts are positively charged thereby bind with the mostly negatively charged colloids.

SAQ 4.4
a) i
b) ii–iv
c) ii–iv
d) i
e) ii–(iii)
f) ii
g) ii and iv
h none
i) iii
j) i, ii and iv (or will accept just i and iv)

SAQ 4.5 Ozone, oxidizes/destroys pesticides.

SAQ 4.6 Chlorine, provides a stable residual for sanitisation of the distribution network.

SAQ 4.7 Oxidation of ammonia to innocuous products (e.g. nitrogen gas and nitrate) via chloramines, such as NH_2Cl and $NHCl_2$.

Solutions to Self Assessment Questions

Unit 5 – Sorption Processes

SAQ 5.1

a) granular activated carbon
b) ion exchange sodium cycle softening
c) granular activated carbon

SAQ 5.2

Typically in the ranges 1.6–2.0 for SAC and 2.3–2.8 for SBA.

SAQ 5.3

Activated carbon adsorption can be used as granular or powder activated carbon for the remove taste and odour, colour, THM precursors and pesticides.

Granular activated carbon can be regenerated by heating in a furnace.

The relative merits of granular and powdered activated carbon are that GAC has high CAPEX and low OPEX costs for large permanent installations whereas PAC low CAPEX high OPEX for small plant/seasonal use.

SAQ 5.4

A two bed ion exchange deionisation process has a strongly acidic cation exchange resin, use either a hydrogen cycle/acid regeneration and removes cations in exchange for H^+ and has a strongly basic anion exchange resin, that uses either a hydroxyl cycle/caustic regeneration and removes anion in exchange for OH^-.

A bed deioniser is and equilibrium process whereas a mixed bed deioniser removes both H^+/OH^- products and hence is not equilibrium so low leakage.

Solutions to Self Assessment Questions

Unit 6 – Biological Processes

SAQ 6.1 Reactor systems used in biological wastewater treatment are continuous and fed-batch reactors.

The different growth mechanisms employed by microbes in wastewater treatment are *suspended growth* processes – microbes are free in suspension and *fixed film* processes – inert surface provides an environment for microbes to attach and grow.

SAQ 6.2 MBRs are biomass separation membrane bioreactors. It is an amalgamation of a suspended growth reactor and a membrane filtration device into a single unit process. Configurations include sidestream and submerged reactors.

SAQ 6.3 In Aerobic processes, microbes use oxygen to maintain biological active, whereas in anaerobic processes microbes utilise alternative electron acceptor (such as nitrate, sulphate and carbon dioxide) to enable microbial metabolism.

Difference between aerobic and anaerobic processes:

	Anaerobic	*Aerobic*
Energy	low	high
% removal	60–90	95+
Sludge production	Low	high
Stability	low-moderate	moderate-high
Start-up	2–4 months	2–4 weeks
Odour	potential	less
Alkalinity	High	low
Biogas production	Yes	no
Nutrients	low	can be high

SAQ 6.4 Activated sludge processes consist of an aeration tank and a settlement tank. The mixed liquor suspended sludge (MLSS) in the aeration tank is maintained by the use of aeration diffusers in the reactor or surface aerators. The waste sludge from the settlement stage is proportional returned to the aeration tank as returned activated sludge (RAS) to maintain biomass or discharged as final effluent.

SAQ 6.5 During anaerobic treatment proteins, carbohydrates and nucleic acids are broken down to long and short chain fatty acids such as butyrate by the hydrolysis or fermentation stage; these molecules are converted to acetic acid during the acetogenesis stage and then final converted into methane and carbon dioxide during the methanogenesis stage.

Fixed-film and suspended growth reactors are used in the anaerobic treatment of wastewater these include, continuously stirred tank reactors (CSTR) such a anaerobic digesters and anaerobic contact reactors, anaerobic filters, anaerobic fluidised bed reactors (AFBR) and upflow anaerobic sludge blanket reactors (UASBR).

SAQ 6.6 Biofilms consist of microorganisms held together in a polymeric matrix attached to a surface. The polymeric matrix consists of EPS, bacteria and inorganic debris.

Biofilm development consists of four phases: (i) the transport and adsorption of organic molecules to a clean surface, resulting in 'conditioned' surface; (ii) transport of cells to the conditioned surface; (iii) growth and multiplication of cells at the surface, development of concentration gradients in the biofilm and (iv) attachment and detachment of cells and other debris to and form the biofilm.

Examples of biofilm processes (i.e. fixed-film processes) include trickling filters, rotating biological contactors (RBC), biological aerated filters (BAF), anaerobic filters, anaerobic fluidised bed reactors (AFBR) and upflow anaerobic sludge blanket reactors (UASBR).

Trickling filters – wastewater percolates down through the medium while air is naturally ventilated up through the filter.

RBC – consist of series of flat or corrugated discs that are half-submerged in the wastewater upon which the biomass grows.

BAF – are submerged, fixed-media reactors with a combine aerobic biotreatment and biomass separation.

Anaerobic filters – are anaerobic reactors that retain the active biomass on static medium.

Anaerobic fluidised bed reactors – consist of a fluidised media that are coated with biomass under anaerobic conditions.

Upflow anaerobic sludge blanket reactors – consist of an upflow biological tank reactor with a combined settling stage and reply on the anaerobic biomass to aggregate into dense flocs or granules.

Solutions to Self Assessment Questions

Unit 7 – Membrane Processes

SAQ 7.1 In dead-end systems the flow is perpendicular to the membrane with the retained particles accumulating on the surface of the membrane "fouling". The thickness of the fouling layer increases over time with the rate of permeation decreasing accordingly. When the filtration rate becomes to low the membrane module or cartridge is replaced or backwashed.

In cross-flow systems the flow in introduced tangentially to the membrane surface which promotes self-cleaning ensuring longer operation periods between replacement and backwashing.

SAQ 7.2 b and d: solute concentration does not effect the change in concentration, and solute (or ion) charge similarly has no direct effect.

Membrane selectivity will have some effect, since CP will change with solute size.

SAQ 7.3 The main membrane design considerations are: a high membrane area to module bulk volume ratio, a high degree of turbulence to promote for mass transfer, a low energy expenditure per unit product water volume and a low cost per unit membrane area and ease of cleaning.

SAQ 7.4 The five principle configurations used in membrane manufacture are: pleated filter cartridge, plate and frame or stack type, spiral wound, tubular and hollow fibre.

Module configuration	Characteristics							
	a	b	c	d	e	f	g	h
Pleated cartridge					X			
Plate and frame			X			X		
Spiral wound			X					X
Tubular	X	X						
Hollow fibre	X			X			X	

SAQ 7.5 Reverse osmosis (RO), Nanofiltration (NF), Ultrafiltration (UF), Microfiltration (MF) and Electrodialysis (ED).

ED, electro-motive force (voltage), IEX membrane, ions (charged species), RO or stripping, concentration differential, dense membrane, gases/VOC.

None.

Solutions to Self Assessment Questions

Unit 8 – Sludge Treatment and Utilisation

SAQ 8.1 Primary sludge from primary treatment; grey and slimy, high odour and more easily digested. Secondary sludge from biological filtration; brown colour, earthy smell and organic matter. Secondary sludge from activated sludge treatment; contains microscopic organisms and biological flocs. Tertiary sludge from processes designed to improve or polish effluent.

SAQ 8.2 Typical treatment system; screening waste to landfill; Thickening of WAS using a belt press and gravity thickening for primary sludge.

Mixing the two sludge types prior to Mesophilic Anaerobic Digestion (MAD) for 12 days for primary digestion and 14 day for secondary digestion if sludge is subsequently used in agriculture.

Alternatively (instead of MAD) dewater and thermal treatment.

SAQ 8.3 Public perception of over health scares; number of pathogens and concentration of heavy metals; land availability; operational spreading difficulties; pollution threat due to surface water run-off; odour emissions, storage – public nuisance and water pollution and transportation costs.

SAQ 8.4 Land applications provide nutrient recycling reduces the need for fertilisers and higher crop yields. Organic matter helps soil fertility, quality and structure. Sludge disposal route is relatively straight forward and cost effective.

Landfilling of sludge is a waste of the resource; more stringent environmental standards for gas and liquid emission we restrict landfill disposal. Landfill tax on waste disposal aims to divert sludge to land and incineration.

Incineration can provide an energy recovery route for sludge. Government grants on renewable power generation and co-gasification and co-combustion for sludge.

SAQ 8.5 The use of digestion pre-treatments such as chemical, mechanical, biological and thermal processes which enhance the hydrolysis breakdown step of complex organic molecules to less complex molecules would improve the digestibility of surplus activated sludge (SAS). Mixing SAS with primary sludge; thickening the SAS prior to digestion and operate the digester correctly at mesophilic temperatures would also improve biological SAS digestion.

Solutions to Self Assessment Questions

Unit 9 – Odour Management

SAQ 9.1 Odour strengths can be determined by the measurement of the odour concentration (OC) by a panel of human assessors using the CEN standard EN 13725 for olfactometry. H_2S measurements can also be used as a surrogate for odour strength. This method is cheaper than olfactometry, but limited due to measuring only a single compound.

Specific compounds (i.e. H_2S) and OC can be used to determine the impact of odorous emissions on a local population. The measurements can be plot to produce a simple contour map of the values or used in dispersion models to predict the dispersion of a gaseous compound from a source to a receptor (i.e. public).

SAQ 9.2 Biological treatment systems utilise microbes on supported media to oxidise odorous compounds that are first dissolved in moisture or liquid surrounding the media. The microbes (bacteria, fungi) then either directly or indirectly oxidise the compounds.

The correct physico-chemical environment is needed to support the growth of a microbial population. Factors that influence microbial growth and therefore performance are: oxygen availability, gas transfer of odorous compounds to liquid, moisture content, temperature, pH, macro and micronutrients in the scrubber liquid.

IWA Publishing's authorised EU representative for General Product Safety Regulations is Diane D'Arras, 15 rue Duret, 75116 Paris, France, e-mail: safety@iwap.co.uk.

Printed and bound by CPI Group (UK) Ltd, Croydon, CR0 4YY
12/05/2026
02108876-0005